Microscale Organic Chemistry

Laboratory Techniques

3rd Edition

NYENTY ARREY

Capital University

Published by Linus Learning.

Ronkonkoma, NY 11779

ISBN 10: 1-60797-556-4

ISBN 13: 978-1-60797-556-4

Printed in the United States of America.

This book is printed on acid-free paper.

Print Number   5   4   3   2   1

Whoso Abideth To Instruction,
Is Willing To Learn: His Or Her
Ability To Understand Will Increase
One Thousand And One Times.

Dr. Nyenty Arrey

The third edition of organic chemistry laboratory Techniques takes the manual to a new direction focusing on what industries and graduate schools are looking for in a recent graduate with a chemistry degree. Three areas of interest are:

1. Chemical Techniques:

    - Extraction
    - Filtration
    - Temperature controlled reactions
    - Evaporation (rotary)
    - Column chromatography
    - Recrystallization
    - TLC
    - Distillation

2. Laboratory Equipment:

    - Fume hood
    - HPLC
    - NMR – Interpretation of spectra
    - PH meter
    - GC/MS – Interpretation of chromatogram/spectra
    - Melting point apparatus
    - IR – Interpretation of spectra
    - UV-Vis

3. Factors in Hiring:

    - Interpersonal skills
    - Team work
    - Undergraduate research
    - Internship experience
    - Grade point average

Students who have used this book give it high praises for enhancing their laboratory skills and they have also become proficient in the interpretation of spectra. The best way to teach and learn spectroscopy is in a laboratory setting.

PREFACE

The purpose of this text is to present the basic techniques used in the organic chemistry laboratory. Understanding and using these techniques will enable students to complete their work correctly and on time the first time it is performed. This text is also based on what industries and graduate schools desire.

This text covers the course materials in depth and is applicable for advance references. You may wish to consult additional references to learn more about approaches to different techniques. It is not claimed to be a comprehensive compilation of information to meet all possible needs and circumstances; rather, the intention is to provide sufficient guidance that will allow students to focus on particular instruments and techniques to carry out experiments under conditions which will offer the highest chance of success.

ACKNOWLEDGEMENTS

I want to express my deep gratitude to the people who were so important and valuable to me in the writing of this new edition. My wife, Regina and kids, Martin, Stephen and Junior for all their support and patience. I'd like to thank my organic chemistry students (past and present) for using the book and making suggestions on how to improve on it.

WHY SINGLE-SIDED PAGES

As an educator, I have intentionally designed this manual to have as many blank pages as possible for student use. The pages could be use during pre-lab discussions and for new ideas or instructions obtained in the laboratory while performing the experiment.

Table of Content

C h a p t e r ①

INTRODUCTION

The organic chemist concerns himself or herself with two classes of organic substances -- those which occur in nature and those which are synthesized in the laboratory. Naturally occurring organic compounds occur as a mixture with other organic or inorganic materials and must be isolated before meaningful identification and study can be accomplished. Although isolation and identification are as old as organic chemistry itself, newer and better techniques are being developed every year. Organic chemists in the 21st century now have many techniques available for the isolation and identification of organic compounds which were unknown a generation ago.

The technique used for isolation of substances from their contaminants depends on the properties of the desired substance as well as the properties of those substances from which it is to be isolated from. If the desired compound is soluble in a solvent and the other material is insoluble, recrystallization may be used. Such is the case in the isolation of acetanilide from sand. The mixture is selectively dissolved in hot water.

In other cases two or more substances may be separated by differences in volatility (i.e. different boiling points). This technique, called distillation (simple or fractional), is used to separate the gaseous, liquid, and solid hydrocarbons from coal tar.

Another important technique for separating mixture is chromatography. There are different types of chromatography, for example column, thin-layer, paper, and gas. Column chromatography involves the differential adsorption of materials on adsorbents such as silica gel, alumina, and others. The mixture is usually added as a solution to a tube containing the adsorbent. By using solvents to elute the column, the substances in the mixture will pass through the column at rates depending upon the degree to which they are adsorbed. Gas chromatography, thin-layer chromatography and paper chromatography are useful for qualitative and quantitative determination of substances.

After a substance has been isolated or synthesized, it is then necessary to identify it. Identification may be established by determining the physical and chemical properties. The physical properties in the case of liquid consist of the boiling point, density, etc. For solid, one of the most important physical properties is the melting point. Chemical properties consist of qualitative and quantitative elemental analyses, structural determination (IR, NMR, MS, etc.), and chemical reactions.

Most of the above isolation and identification techniques will be used in this course.

Safety

Working in an organic chemistry laboratory poses certain risks which we have attempted to minimize. However, safety in the laboratory requires a strong commitment from everyone involved. The safety video "Working in a Chemistry Lab" should be watched by the instructor, teaching assistants and students before performing the first experiment.

1. Safety goggles must be worn at all times. Do not wear contact lenses in the lab.

2. Sandals should not be worn in the lab at any time and shorts are discouraged. Clothing is much easier to replace than skin!

3. Never work alone in the laboratory. Someone must always be aware of what you are doing. Unauthorized experiments are not permitted.

4. Know the location and purpose of the safety devices in your lab.

5. Dispose of glassware and chemical wastes in the containers provided. Minimize the amounts of chemicals you use.

6. Know the properties of the chemicals you are using.

7. Avoid the use of open flames.

8. Avoid contact with the materials you are handling. Inhalation and absorption through the skin or open cuts are common routes of entry. Gloves may be recommended for certain operations.

Location of Safety Devices in the Laboratory

Laboratory Hygiene

It is important to leave the lab clean for the next group of students. Spills that you make around the balances and side shelves should be cleaned up immediately. Appropriately dispose of used pipets, weighing papers, *etc.* Before you leave the laboratory, gather up your equipment, sponge off your work area, and lock your locker.

Sources For Organic Chemical Information

1.   www.chemexper.com

2.   www.chemfinder.com

3.   www.msds.com

4.   MSDS Binders in the labs

5.   Aldrich Chemical Book of Reference or CRC Handbook

6.   www.chemSpider.com

7.   W. Chemistry Handbook

C h a p t e r 2

MEASUREMENT OF WEIGHT

To obtain good results in organic chemistry experiments, the chemist must have the ability to measure reagents (solids and liquids) accurately. Solids to be used in an experiment will be placed next to the balance in the laboratory while the liquids will be in the designated hoods.

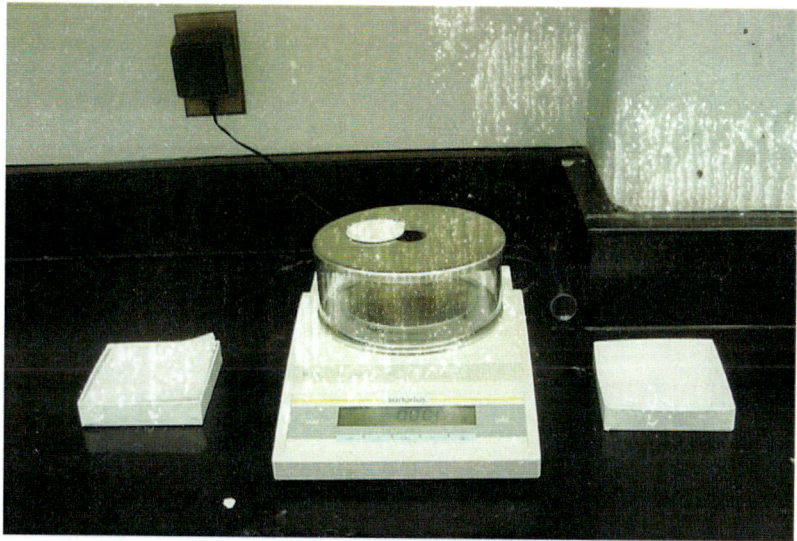

Top-loading balance

Procedure: Using Top-Loading Balance

Solids

1. Turn the balance on if it is off.

2. Place a weighing paper on the balance pan.

3. Press the **Tare** key so that the paper appears to have zero weight.

4. Add your solid using a spatula until the balance gives the desired weight.

5. Transfer the weighed solid to the desired container.

6. Close the reagent container and clean any spills.

Liquids

1.   Place the flask or vial on the balance.

2.   Press the **Tare** key to show zero weight.

3.   Transfer the liquid from the reagent bottle to the flask or vial using a pipet.

4.   Stop adding when the desired weight is obtained.

5.   Close the regent bottle and clean any spills.

Note:

If you know the density of the liquid, you can calculate the volume and then use a syringe to deliver the desired amount.

$$\text{Volume(ml)} = \frac{\text{Weight(g)}}{\text{Density(g/ml)}}$$

LABORATORY NOTEBOOK

Keeping records of laboratory work is very important. If an experiment is performed and there is no record of what was done, the result will be questionable and the chemical community will not accept the results. Chemists in different institutions will develop their own style for recording experimental data, but there is certain information that will be common in their records. The laboratory notebook should be a diary of experiments performed and should contain exact details of how experiments were carried out. There are different types of laboratory notebooks, and as such, institutions make their decision on what type to use. Laboratory notebooks don't have to be very neat, but should be legible. If you make a mistake while writing, just draw a line through it and move on.

What is the purpose of a laboratory notebook?

There are so many reasons why we should have a laboratory notebook:

1. It tells us the exact procedure that was used in an experiment, and as such, we can refer to it in case of any problem. This can be very important especially if the experiment was not successful.

2. It is the main source of reference when you want to write reports, papers, etc.

3. It tells us when an experiment was carried out.

4. It enables another student to follow your work and be able to repeat the experiment with little or no problem.

NOTEBOOK FORMAT

There are different formats for record keeping in the laboratory, but there are some essential features that should be included in any format. For example:

1. Experiment number

2. The date

3. Experiment title

4. Table of physical properties of reagents to be used in the experiment

5. A reaction scheme indicating the proposed transformation (Chemical Equation)

6.   The procedure.

7.   Diagrams of equipment to be used.

8.   Observations/Data collected.

9.   Conclusion.

In this book, two formats will be addressed: 1) Guided Inquiry-Based format and 2) Synthetic chemistry experimental format. In both formats, I strongly encourage students to write features 1-6 in their laboratory notebook before coming to lab. There are some advantages associated with this suggestion:

1.   Students read the experiment before class.

2.   Students are mentally prepared to work.

3.   They ask questions during pre lab session for clarification.

4.   They don't spend unnecessary time in the laboratory.

Example of both formats follow:

Guided Inquiry-Based Format

Sample Distillation-ID of Unknown

Purpose

The purpose of this lab is to use simple distillation to purify an unidentified organic liquid, then use gas chromatography and boiling point techniques to identify the liquid.

Compound	Structure	Molecular weight	Boiling point (^{0}C)	Melting point (^{0}C)	Density	Toxicity	Solubility
methanol		32.04	64.7	-98	0.791	flammable liquid poison	miscible
1- propanol	OH	60.10	97.2	-126	0.803	colorless liquid w/ non-receival alcohol	miscible
2- propanol	OH	60.10	82.4 80.9	-88.5	0.785	clear, color-less liquid w/ an order rub-ber alcohol	miscible
butanone	O	72.11	79.6	-86.3	0.805	flammable liquid	Sol. (25.6g/100ml at 20^{0}C)
hexane		86.18	69	-95	0.655	flammable, colorless w/a mild gaso-line-like odor	slightly sol. 1000947g/ 100ml
ethyl acetate	O O	88.11	77.1	-83.6	0.895	flammable, colorless w/a pleasant, fruity odor	Mediately 8g/100ml
cyclohexane		84.16	77.1	6.6	0.779	flammable, colorless w/a mild, sweat odor	Slightly <0.1g 1100ml at 17^{0}C

Procedure

Setup using Hickman Still

1. Put sand in an evaporating dish and place it on a hot plate.

2. Put 2.0 mL of unknown liquid and spine vane into a 5.0 mL conical vial.

3. Put the conical vial into the sand.

4. Put the Hickman Still on the vial.

5. Clamp the Hickman Still at the top with a 3-finger clamp.

6. Set the hot plate to a medium setting.

7. When the liquid starts to boil, look for liquid starting to collect in the collar of the Still. May want to place a short disposable pipet down the throat of the Still to see if any liquid can be drawn off.

8. Transfer (use short pipet) the liquid from the top of the Still into a clean, labelled sample vial.

9. Continue transferring liquid to the sample vial until liquid in the reaction vial drops below 0.5 mL. Then stop the distillation.

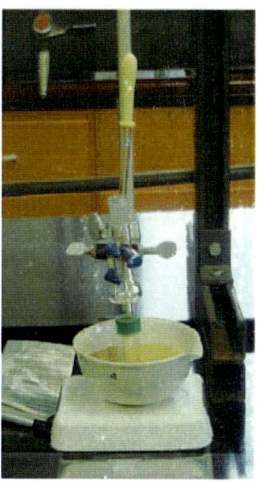

Distillation Set-up

Boiling Point

1. Prepare a sand bath.

2. Place about 0.5 – 1.0 mL of the distilled unknown along with a boiling chip into the test tube.

3. Clamp the thermometer so that the probe doesn't touch the sides of the test tube and the tip of the probe is above the liquid.

4. Heat the system until the liquid starts to boil.

5. When the thermometer is at the highest and stays there for about one minute, record the temperature (This is the boiling point of the liquid).

Gas Chromatograph

1. Use a microliter syringe to inject the sample of unknown liquid into the gas stream of the GOW-Mac instrument through the injection port and into the column.

2. As the sample hits the detection, the detection generators an electric signal.

3. The instrument will read the signal. Then interpret the GC sample through the column..

4. Analyzer measure the retention time (t_R), which is the time it takes the sample to min through the column.

    *   The air peak in the chart paper is the reference point that marks the start. If there is no air peak, make a mark on the chart paper as soon as the sample has been injected.

    *   Find the distance the chart paper crawls in a minute. Measure the distance from the starting point to the mid point of each peak on the baseline.

Example:

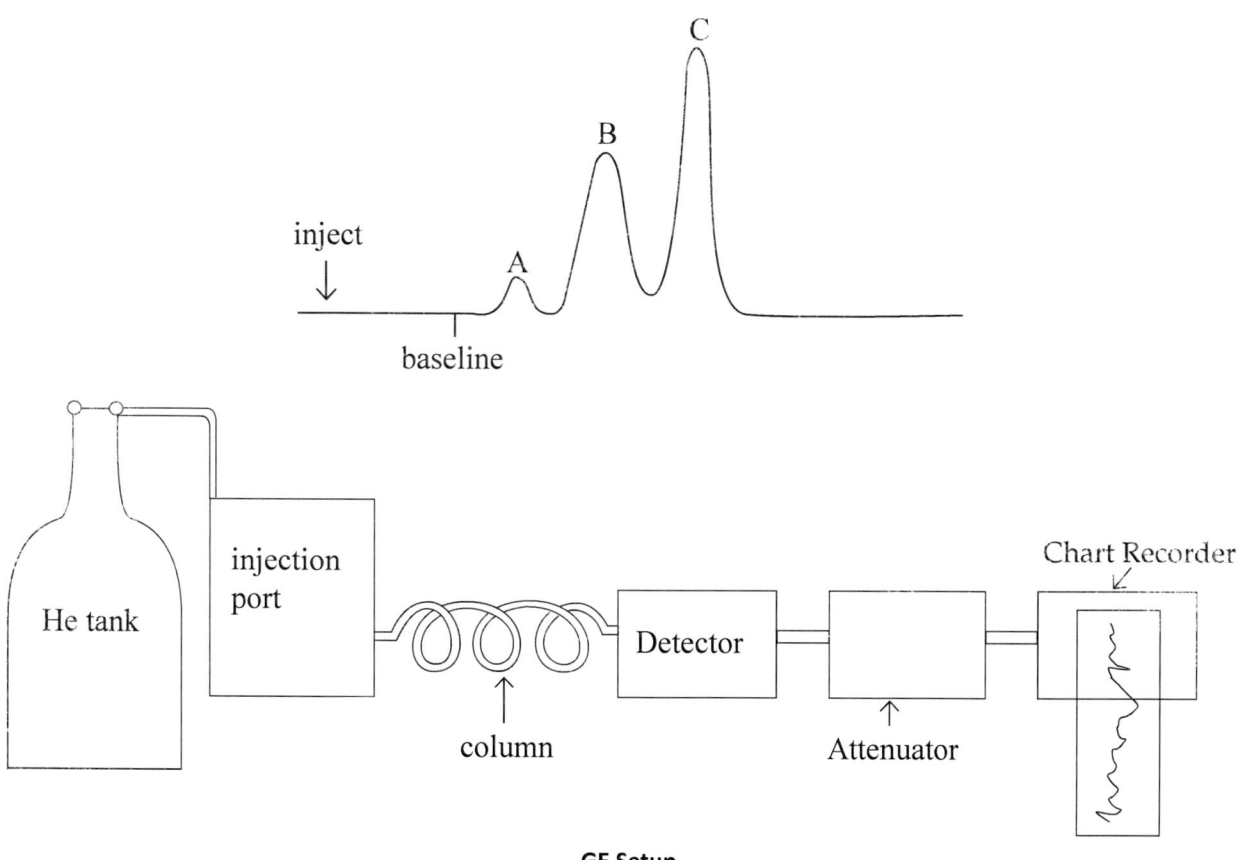

GE Setup

Synthetic Chemistry Format

Purpose

The purpose of this experiment is to prepare benzocainefrom reacting P-amainobenzoic acid w/ethanol. Then analyze the product Using IR, 'HNMR and melting pomx.

Chemical Equation

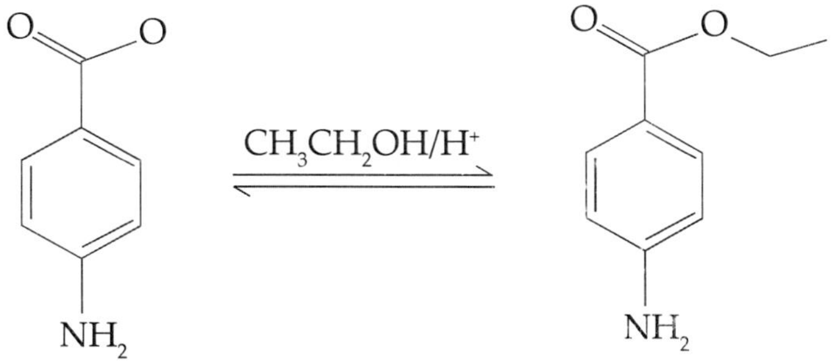

$$CH_3CH_2OH/H^+$$

Theoretical Yield

P-amminobenzoic acid

MW = 137.15g

Used = 0.12g

Mol = 8.75x10⁻⁴

4- aminobenzoic acid ethyl ester

MW = 165.2g

Used = 0.145g

mol = 8.75x10⁻⁴mol

Tables

Structure	MW	Density	M.P(°C)	B.P(°C)	Toxicity	Solubility
p-aminobenzoic acid (O, OH / NH₂)	137.15	1.37	186.189	-	harmful	
ethyl p-aminobenzoate (O, O— / NH₂)	165.19	-	89	172 at 17mm µg	-	-
ethanol (OH)	46.07	0.789	-114.1	78.3	flammable	misable
Na⁺O⁻ O⁻Na⁺ (carbonate)	105.98	851	1600	irritant	-	2.53
HO—S—OH (O, O)	98.07	3	280	corrosion	soluble	1.84
—OH	32.04	-98	64.6	Poison Hammable	miscible	0.791

Procedure

1.   Place 0.12 g of p-aminobenzoic acid and 1.2ml of ethanol in a 5 ml conical vial.

2.   Add spine vane and stir mixture until solid dissolves completely.

3.   While stirring add 0.10 ml sulfuric acid. Then attach air condenser.

4.   Heat the mixture for 45 mins to 60 mins until solution is clear.

5.   Remove from heat and cool to room temperature.

6.   Transfer contents into a small beaker with 3.0 mL of distilled water.

7.   Add 1.0 mL of 10 % sodium carbonate and stir.

8.   Check pH when no more gas forms.

9.   Collect benzocaine by suction filtration.

10.   Wash benzocaine 3 times with 1 ml of distilled water.

11.   Allow to dry overnight and then perform tests.

12.   Place crude product into Craig tube and add minimum amount of methanol/water mixture.

13. Heat until solid dissolves.

14. Allow to cool to room temperature then ice-bath for crystals to form.

15. Insert inner plug in to tube and collect crystals by centrifuge.

16. Weigh product and perform m.p., and IR.

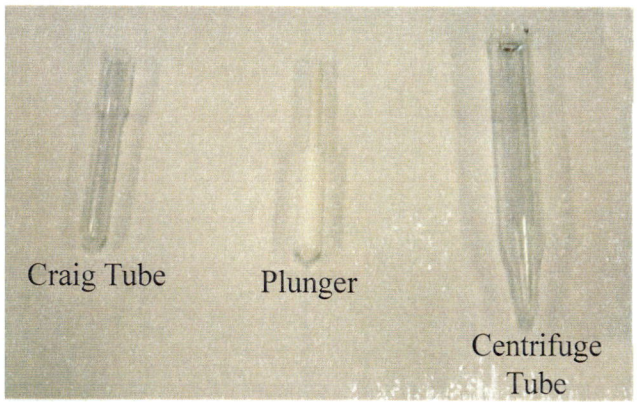

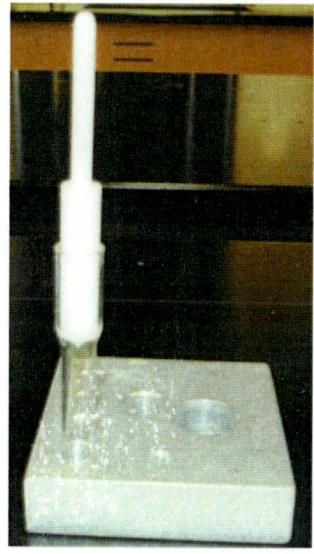

Plunger in Craig Tube

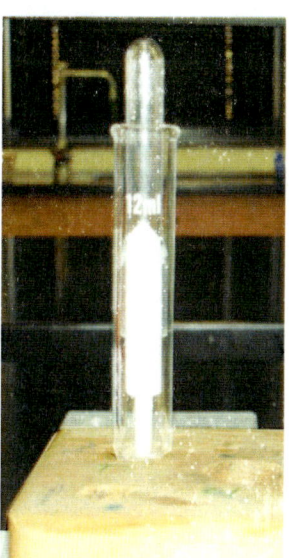

Craig Tube and Plunger
in Centrifuge Tube

MELTING POINT

The melting point is one of the simplest and yet most important methods for identifying organic solids and determining the purity of substances. Most organic solids melt in the convenient range of 50-250 °C. his is in striking contrast to inorganic salts which, for the most part, melt at much higher temperatures. The melting point is defined as the temperature at which a solid changes to a liquid. From a thermodynamic point of view, the melting point is the temperature at which a solid and its liquid have equal vapor pressures.

Pure organic solids will usually melt very sharply. Due to the fact that most laboratory melting point apparatuses have less than perfect heat transfer, a melting range rather than a point is obtained even for pure substances. **However, the range should be no greater than 2 °C for a pure substance even with the apparatus used in this course.** Whereas a pure substance melts sharply, a mixture of substances will melt over a wide range. This is true even though the components of the mixture individually melt at the same temperature. For example, both urea and cinnamic acid melt at 133 °C. A mixture of the two may begin melting as low as 100 °C and melt over a 15-20 °C range. A general rule may be formulated that a mixture of two substances will melt lower than either of the two substances alone.

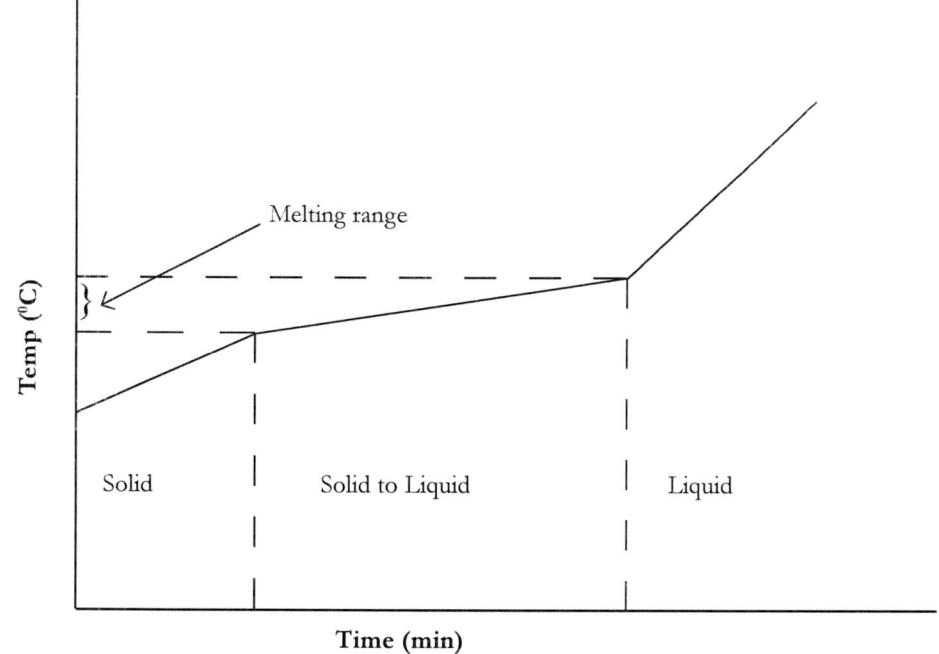

Melting Point

As can be seen from the above discussion, the identity of a substance cannot be made on melting point alone, in as much as there may be several organic compounds with identical melting points. A mixture melting point, however, will establish the identity or non-identity of two similarly melting components. There are a few examples known in which two substances may be mixed without lowering the m.p. But these are rare and will not be encountered in this course.

Procedure

To take a melting point (m.p.) in the laboratory, various kinds of m.p. apparatus may be used. These include a Thiele tube containing mineral or silicone oil which is convection-stirred as the tube is heated, a small beaker containing oil which is hand-stirred by a circular glass stirrer, an electrically-stirred and heated oil bath, or a metal block which is electrically heated. Inaccurate m.p.'s will be obtained if the bath is heated too rapidly because the sample may not have had time to equilibrate with the oil bath. A general rule is to heat the bath rapidly to approximately 15° below the m.p. and then heat so that the bath rises 1-2 °C per minute. **If the m.p. is not known, it will be necessary to get an approximate m.p. by heating a sample rapidly.** Then, the more accurate m.p. will be obtained by the method above using a new sample. In this laboratory, we are going to use a metal block which is electrically heated.

The sample is placed in a thin capillary tube of approximately 1 mm inside diameter and 75-80 mm long. The tubes may be prepared by heating a piece of 10 mm soft glass tubing with the Bunsen burner until the heated portion is very soft and then pulling the ends of the tube apart slowly. A piece of 10 mm tubing 25 cm long should be used so that the hands will not be burned. Cut the capillary tubing into 75-80 mm lengths with a sharp file, carboundum chip, or sharp-edged piece of porcelain. Seal off one end of the capillary with a burner.

To add the sample to the capillary tube use the following procedure :

Crush the sample into a fine powder on a watch glass or weighing paper by means of a spatula and make it into a small "pile". Force the open end of the capillary tube into the pile of solid. The small column of solid, which has entered the open end of the capillary, may be shaken to the bottom of the capillary by gently tapping the closed end on the desktop. Alternatively the capillary may be dropped through a 25-30 cm long piece of glass tubing. A sample of about 1-2 mm high in the capillary tube is sufficient for a melting point.

Empty capillary tube, Weighing paper, Watch glass, and Capillary tube with sample.

In recording melting point ranges, record the temperature at which the first liquid is observed and the temperature at which the last solid changes to liquid. For example, it is rare that a melting point of 120 °C is observed; it is more common to observe m.p. of 120-121 °C.

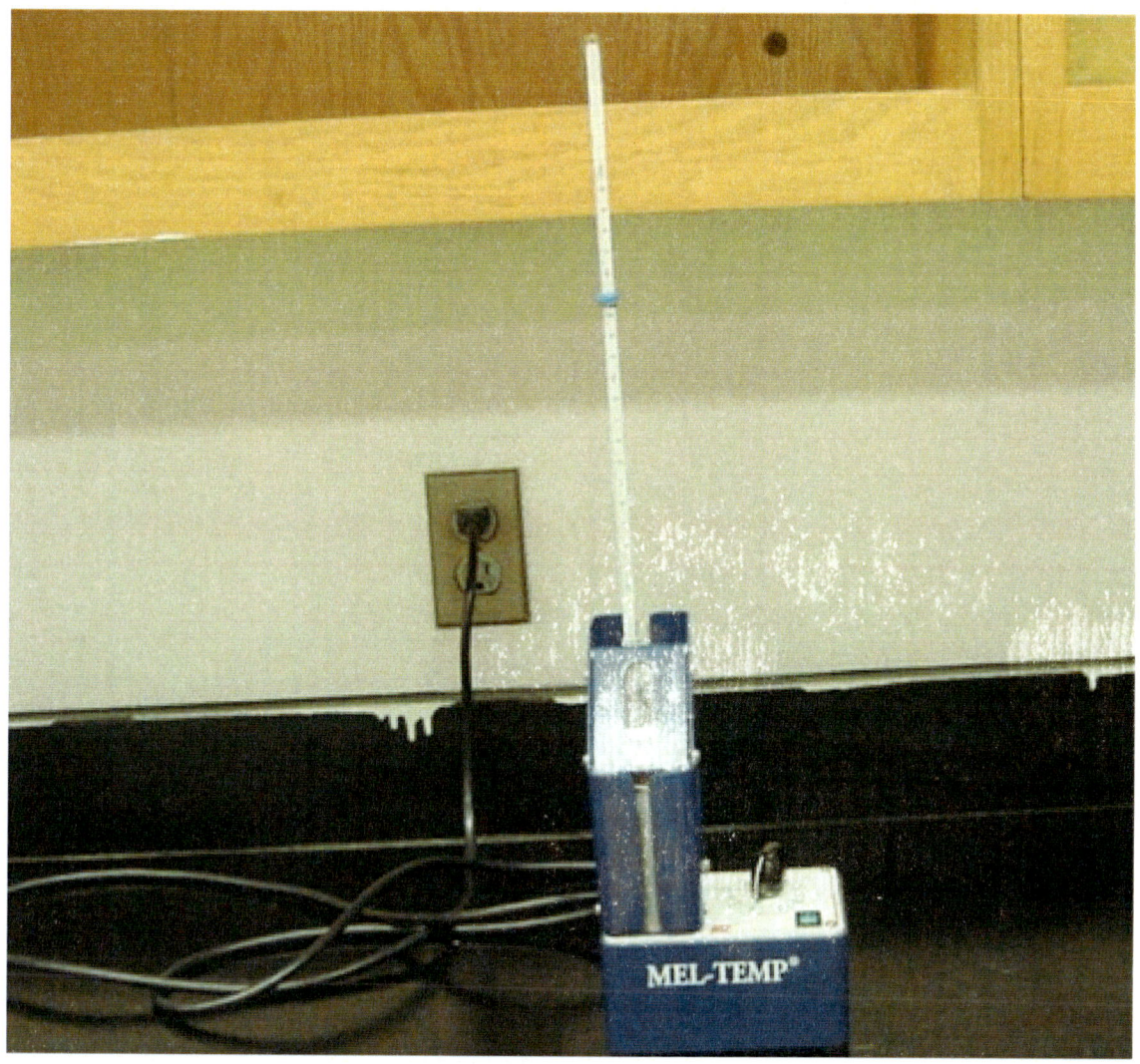

Melting point apparatus

C hapter 5

DISTILLATION AND BOILING POINTS

Distillation may be defined as the conversion of a liquid to its vapor followed by condensation to the liquid state. Distillation is valuable in separating substances of different volatilities. While the molecules in a liquid are not held tightly enough together to form a definite crystalline pattern, there are attractive forces such as van der Waal forces, hydrogen bonding, etc., which must be overcome before the liquid will vaporize.

A liquid (or for that matter, a solid) exerts a vapor pressure at any temperature. If a liquid is placed inside a closed container, molecules will escape as gas from the surface of the liquid and gas molecules will reenter the liquid. Eventually the rate of vaporization will equal the rate of condensation. The gaseous molecules above the liquid exert a pressure. The vapor pressure rises with temperature and may be measured by connecting the container to a U-tube, which contains a column of mercury.

When a liquid is heated until its vapor pressure equals the external pressure, the liquid boils. The boiling point, defined as the temperature at which the v.p. of a substance is equal to the external or applied pressure, may be lowered or raised by changing the external pressure. For example, in a pressure cooker where the pressure is greater than one atmosphere, water boils above 100 oC. On the other hand, substances, which are virtually involatile at atmospheric pressure, may be distilled in a vacuum.

When a pure liquid is distilled the b.p. should remain constant throughout the distillation since the composition of the vapor above the liquid will always be the same as that of the liquid in the distillation apparatus. The b.p. may, therefore, be determined by a simple distillation.

The situation is different however when a homogeneous solution of two or more liquids is distilled. Raoult's law states that the partial v.p. of a volatile component in an ideal solution is equal to its v.p. times its mole fraction of the solution. The total vapor pressure of a solution is equal to the summation of the partial v.p.'s of all the components.

$$P = P_A N_A + P_B N_B + P_C N_C + ... + ... + P_N N_N$$

A solution will begin to boil when the total partial pressures are equal to the external pressure. If there are two liquids in a solution which boil at the same temperature, it can be readily seen that the boiling point of the solution will be constant and at the same temperature as that of each component. Furthermore, there will be no separation of the two since the composition of the vapor will always be the same as that of the liquid.

Consider another solution in which the components have different volatilities or different vapor pressures. The vapor above the boiling solution will be richer in the substance which has the higher vapor pressure. Furthermore, it may be seen by a simple calculation that boiling will not begin until the temperature of the solution is *above* the b.p. of the most volatile component but below that of the less volatile

component. If a simple distilling head is used, it is found that the more volatile component predominates in the distillate at first but the ratio:

$$\frac{\text{More Volatile}}{\text{Less Volatile}}$$

decreases as distillation continues. Furthermore, the boiling point gradually increases toward that of the higher boiling component as distillation continues. However, it should be pointed out that a simple distilling head is useful for distillations only when pure liquids are being distilled, when a liquid is being distilled from a non-volatile solid or non-volatile liquid, or when substances have boiling points which differ by about 40°C or more.

Many types of apparatus are used in distillation varying all the way from simple distilling heads to columns with many theoretical plates. Accurate boiling temperatures are obtained only if there is liquid vapor equilibrium at the bulb of the thermometer used for recording the boiling temperature. This is usually accomplished during distillation if the rate is 1-2 drops of distillate per second. For atmospheric pressure distillations (or refluxing) a boiling "chip" is added to the distillation flask to provide smooth boiling.

Many liquids, which have high boiling temperatures at atmospheric pressure, decompose at those temperatures. Consequently, it is necessary to distill them under reduced pressure. It is advisable to place a shield in front of the apparatus during such distillation because an implosion may take place if there is a flow in any part of the glassware. To obtain smooth boiling during reduced pressure distillation, one of several techniques may be employed. If the substance is insensitive to air, air may be pulled through a fine capillary through the liquid. If air oxidizes the material, nitrogen may be substituted and it is convenient to fill a balloon with nitrogen and fasten it to the tube leading to the capillary. Alternatively to bubbling a gas through the liquid a boiling stick may be placed in the flask. In predicting boiling points at lower pressures, one may use various temperature-vapor pressure curves for known compounds and make the assumption that similar type compounds will behave in a like fashion.

One rule of thumb that gives an approximate boiling point is useful if one knows the boiling point for at least one pressure. In lowering the pressure from atmospheric to 25 mm, the boiling temperature is halved. Below 25 mm, the b.p. is lowered 10°C each time the pressure is halved.

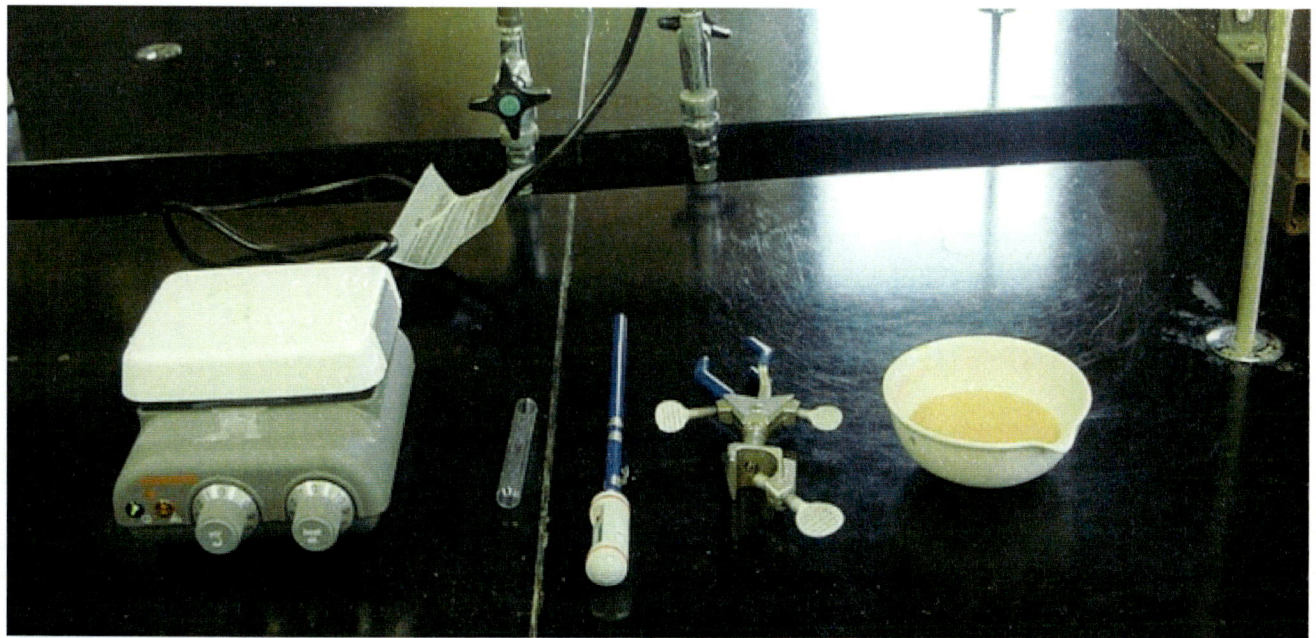

Materials for Boiling Point

Diagrams for Boiling Point

Set-up Overview

Tip of Thermometer above Solution

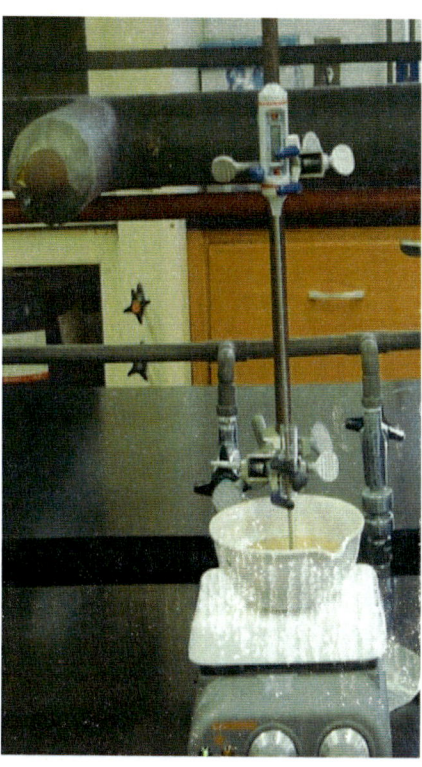

Boiling in Progress

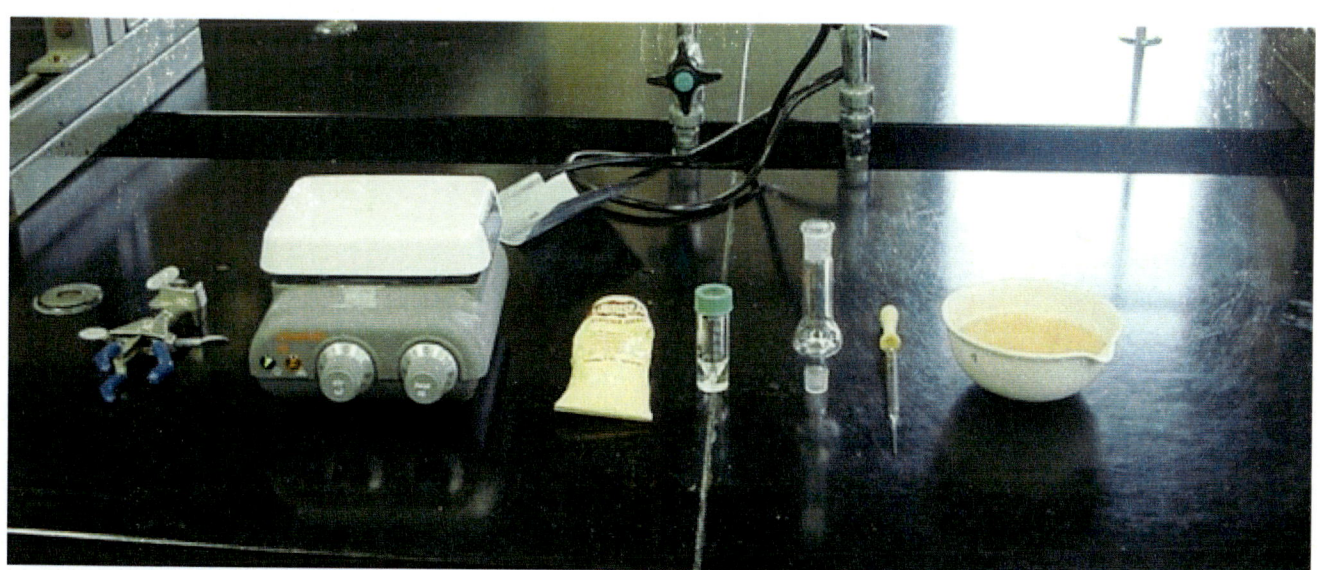

Diagrams for Distillation

Materials for Distillation

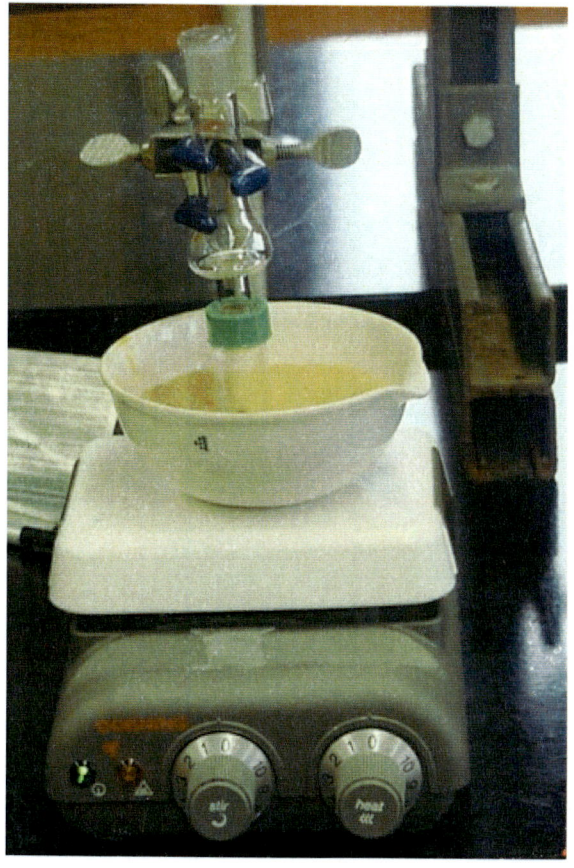

Distillation in Progress

Collecting Distillate

Fractional Distillation

When two liquids have vapor pressures that are close together, it is necessary to use a fractionating column to separate them. This may be a bubble cap type, which is often used in industry, or a tube packed with glass beads, helices or other inert packing. What occurs in a fractionating column is a series of vaporizations and condensations. The liquid in the flask boils. The vapor is richer in the more volatile component. In the column the vapor condenses and as the condensate falls it is revaporized by hot vapor coming up through the column. In this second vaporization, the vapor is still richer in the more volatile components.

The efficiency of a distilling column is expressed in terms of the number of theoretical plates. A column in which the vapor of the distillate is in equilibrium with the original solution has one theoretical plate. The number of theoretical plates will depend upon the number of vaporizations and subsequent condensations in the column and the equilibria between vapor and liquid of the different compositions. The scheme below illustrates a typical vapor-liquid composition curve for a two-component solution consisting of liquids A and B.

When a 50:50 mixture of A and B is boiled, the vapor above the solution will have composition x. A column that will concentrate a composition of 50:50 to x has one theoretical plate. If the distillate contains the composition y, the column has two theoretical plates, etc. Examples of distillation curves are shown. The type of curve obtained will vary some where between 3 and 4 depending upon the efficiency of the column and the difference between vapor pressures of the two components.

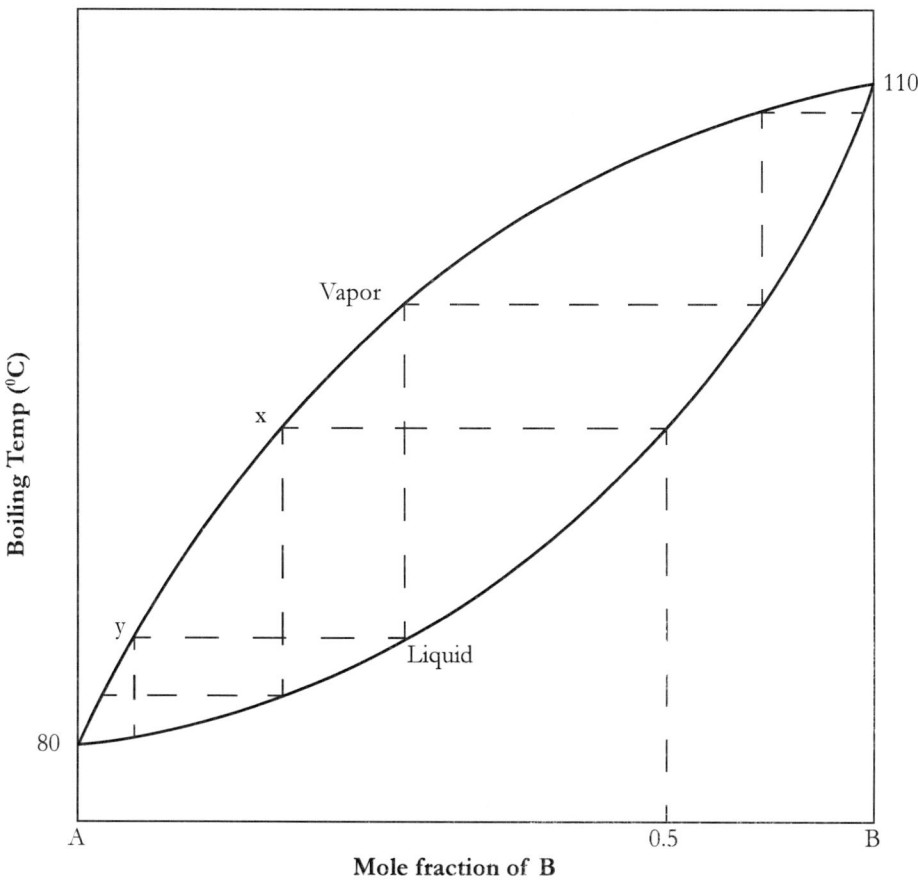

Vapor-Liquid Composition Curve

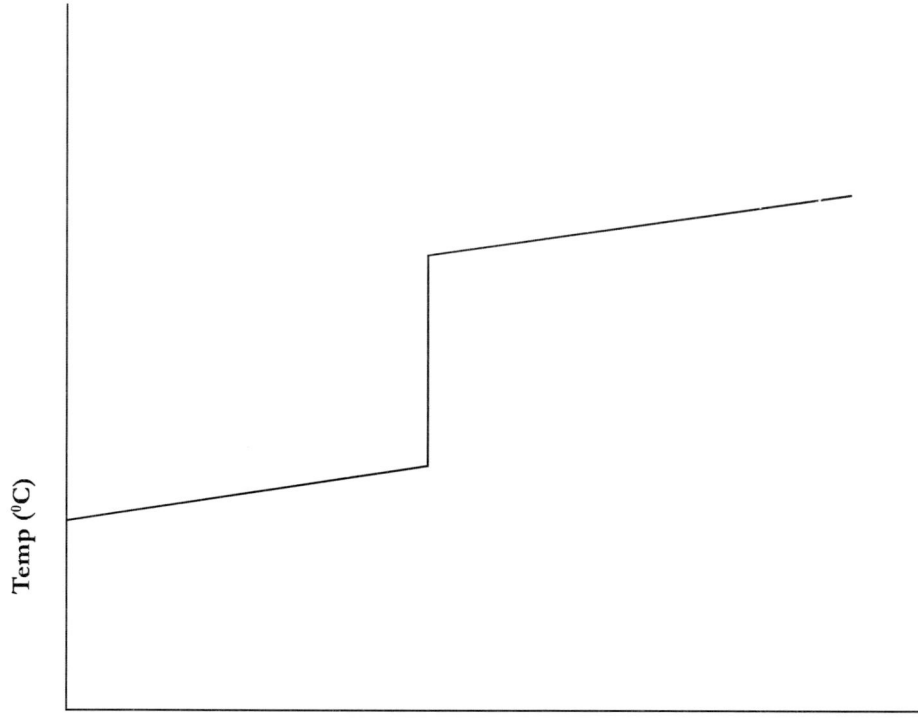

Ideal Curve

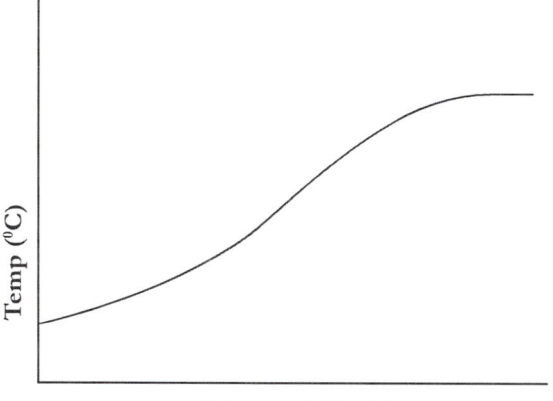

Curve obtain with a simple distilling apparatus

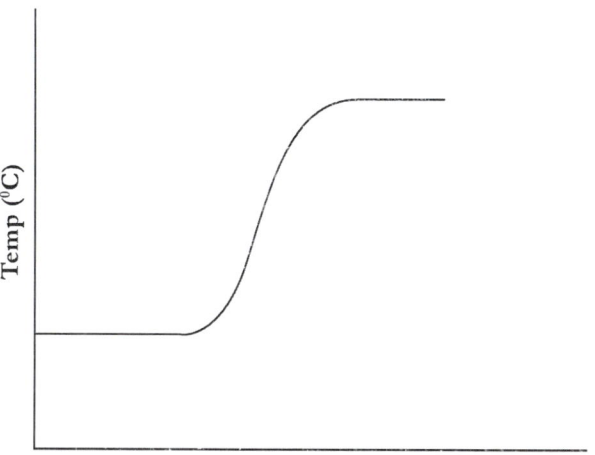

Curve obtained with an efficient distillating column

Rotavapor Distillation

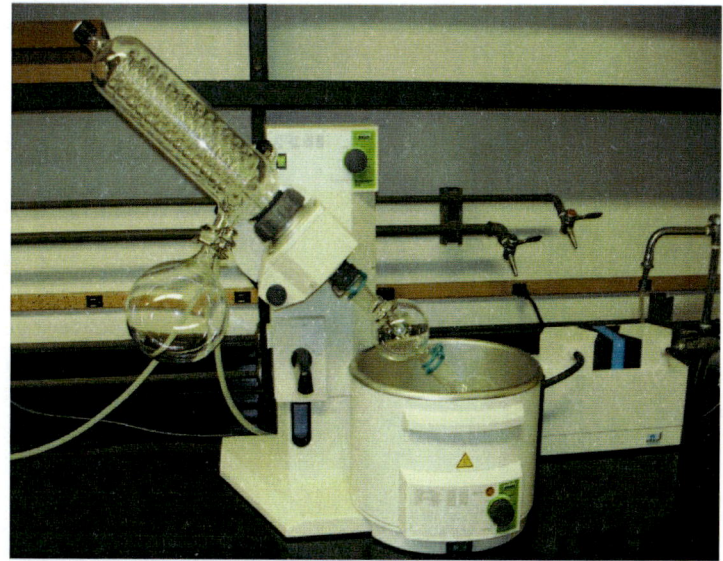

Operation of the Ratavapor R II: The "Ratovap"

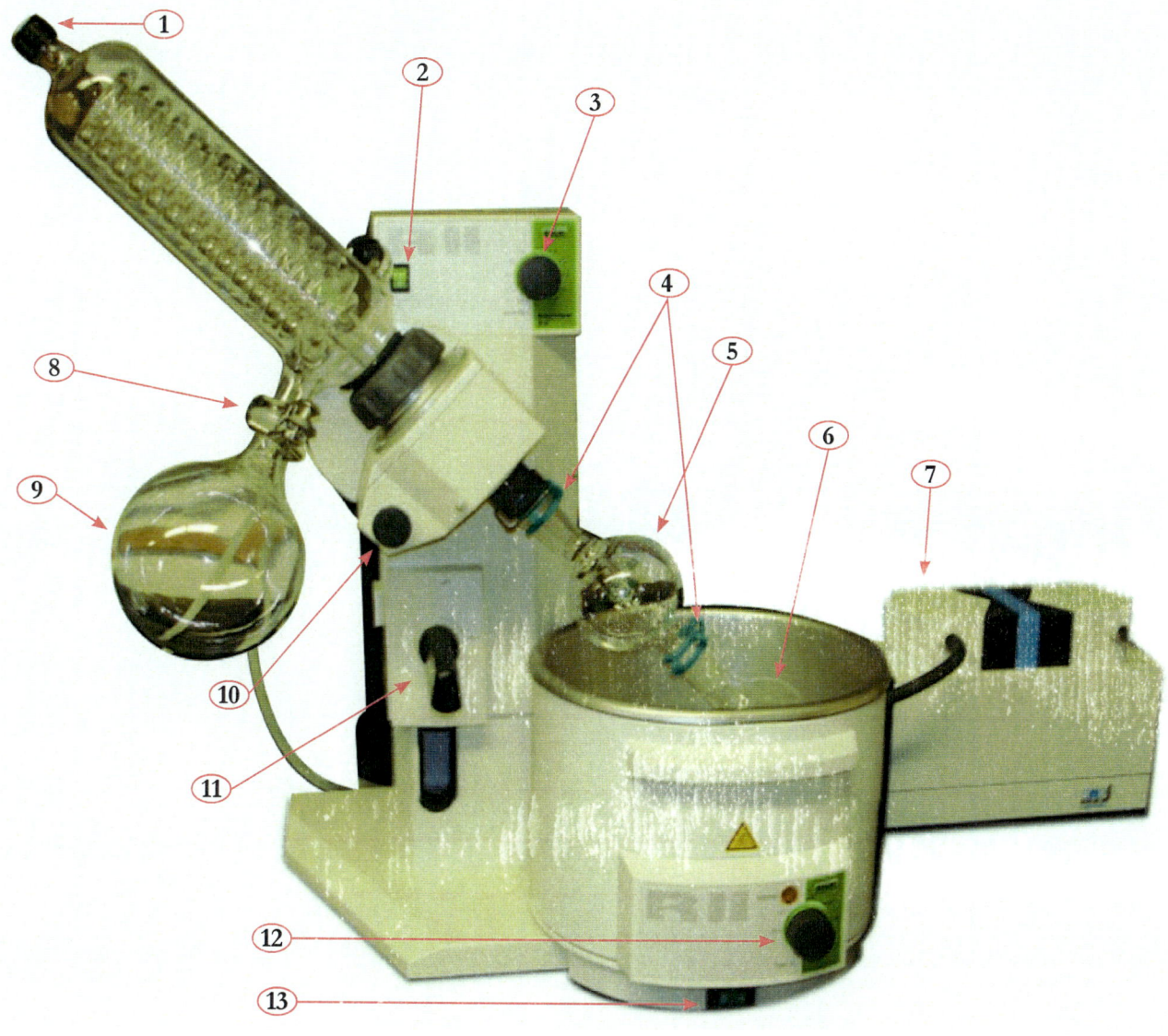

1.	Screw Cap	2.	Main switch for Rotavapor
3.	Knob for rotation speed	4.	Flask clips
5.	Bump trap	6.	Evaporating flask
7.	Vacuum pump	8.	Flask clip
9.	Receiving flask	10.	Knob for immersion angle adjustment
11.	Quick-action jack to raise and lower evaporating flask	12.	Adjusting knob for the heating bath temperature
13.	Main switch for heating bath		

Procedure

1. Check receiving flask, bump trap, condenser and evaporating flask. If they contain any liquid, empty it into organic waste container. If the bump trap is dirty, remove it and rinse it out with acetone.

2. Put your solution into the evaporating flask (NO MORE THAN HALF FULL), apply a small amount of grease to ground glass joint, slide on the evaporating flask and secure with a flask clip.

3. Turn the cooling water on to the condenser coils – a gentle flow is sufficient.

4. Turn on vacuum pump – make sure the screw cap is tight.

5. Lower the evaporating flask into water bath using the Quick-action jack, adjust the speed of rotation using knob for rotation speed.

6. Turn on the water bath and adjust heat accordingly – remember that the apparatus is under vacuum so the water bath doesn't need to be set at the normal boiling point of the solvent, 40-50 °C is usually sufficient.

7. When the distillation is finished, stop rotating evaporating flask, raise the flask out of the water bath, turn off vacuum pump and the water supply.

8. If you don't intend to immediately perform another distillation, turn off the heating bath to save energy.

9. Loosen the screw cap to release the vacuum pressure to be able to access the receiving flask and evaporating flask safely.

10. Remove the evaporating flask for further use or analyses.

11. Transfer liquid in receiving flask into organic waste container.

12. Clean all glassware especially the bump trap when you are done.

13. Make sure everything on the Rotavapor is completely off and the area is clean.

Chapter 6

EXTRACTION

Extraction is a useful technique for the separation of a component from a mixture by means of a solvent. There are two types of extractions that we might encounter in this course **1) liquid-liquid extraction:** involves two liquids that are not miscible. For example, extraction of propionic acid from water into ether. **2) solid-liquid extraction:** for example, extracting caffeine from tea by water.

1. Liquid and liquid

The separation of two immiscible liquids in the laboratory may be best accomplished by adding the mixture to a conical vial or separatory funnel, allowing the two layers to separate, and drawing off the upper layer or the bottom heavier liquid. This technique is also used to extract a substance from one liquid into another immiscible liquid. For example, caffeine is soluble in water but it is more soluble in chloroform. Since chloroform and water are immiscible the aqueous solution of caffeine is placed in a conical vial, chloroform is added, and the vial shaken. During the shaking the vial is occasionally vented to relieve pressure. The vial is then set on the bench top for the layers to separate. The lower chloroform layer or the upper layer is drawn off using a pipet. It is usually necessary to extract with several portions of solvent in order to remove most of the desired material.

Before Extraction

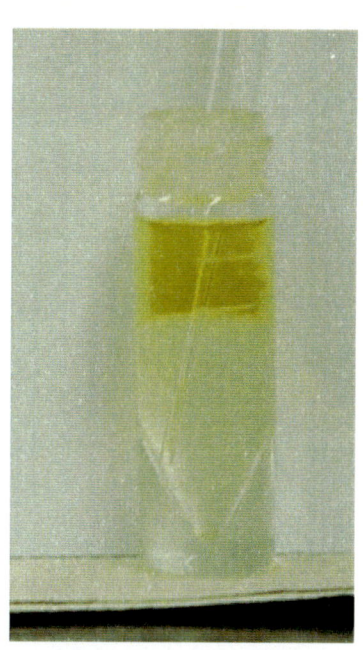

Pipet in the Aqueous layer

Extraction in Progress

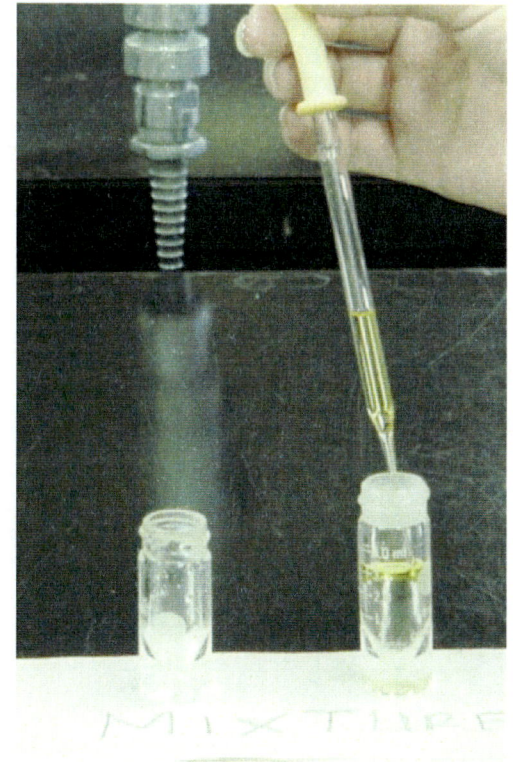

Organic layer in pipet

End of Extraction

Water and an organic liquid are the immiscible liquids most often encountered. After separation, it is usually necessary to dry the organic layer before further work. This is done by adding anhydrous magnesium sulfate ($MgSO_4$), or sodium sulfate (Na_2SO_4) or calcium chloride ($CaCl_2$) to the organic liquid, swirling to allow drying to be complete, and then decanting the liquid to a dry container or filtering off the drying agent. It should be pointed out that $CaCl_2$ should not be used when there are alcohols or amines present since these forms complex with $CaCl_2$.

Properties of a Good Solvent For Liquid-Liquid Extraction

1.   Immiscible with or sparingly soluble in liquid from which solute is to be extracted.

2.   Extracts little or none of the impurities.

3.   Readily dissolves substance to be extracted.

4.   Does not react with solute in an undesirable manner.

5.   Capable of being easily removed from solute after extraction.

2. Solid and liquid

This type of separation is encountered when a solid is recrystallized from a liquid solvent. The filtration may be carried out by gravity filtration by folding a piece of filter paper, placing it in a filter funnel and pouring the solid-liquid mixture onto the paper. Oftentimes such a filtration proceeds too slowly and it is more convenient to use suction filtration. For this type a Buchner funnel is used. Cut a circle from filter paper and rest it on the Buchner funnel (make sure the filter paper is large enough to cover the holes) that is fitted to a suction flask by means of a rubber sleeve. The suction flask is connected to the aspirator (fig. A) or vacuum pump (fig. B) as shown below. Put a 400 mL beaker under the aspirator to control the water flowing from it. Using the vacuum pump is more efficient and it saves water.

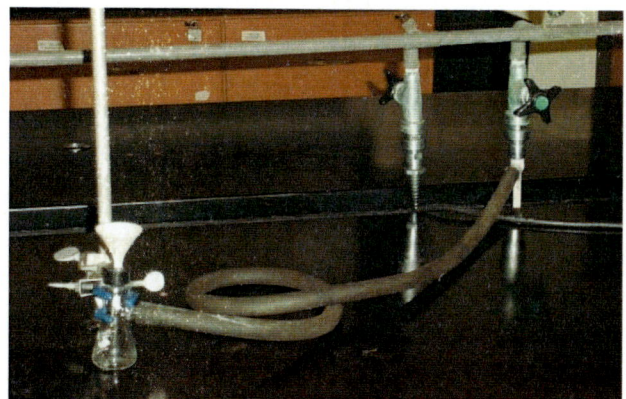 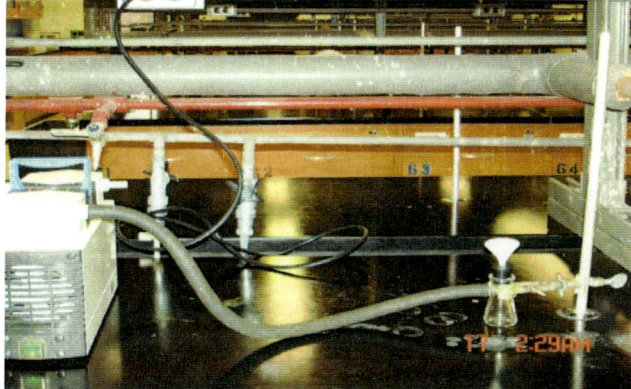

Fig. A   Fig. B

Set–up for Vacuum (suction) Filtration

In carrying out a filtration the mixture of liquid and solid is poured onto the filter, the last traces of crystals are washed onto the filter paper with a small amount of filtrate, and then the solid is washed with a small amount of fresh **cold** solvent. In suction filtration, the crystals are mostly dried on the filter paper during suction by pressing them with a metal spatula. In either type of filtration, the crystals are scraped from the filter paper onto a clean, tared watch glass to finish the drying. To avoid collection of dust on the crystals and also their blowing away, a second watch glass is placed over the one containing the crystals, using small slits of cork between the watch glasses to allow ventilation.

3. Acid-Base Extraction:

Separating a mixture to various components is based on acid-base chemistry. The flow chart below is an example of the process.

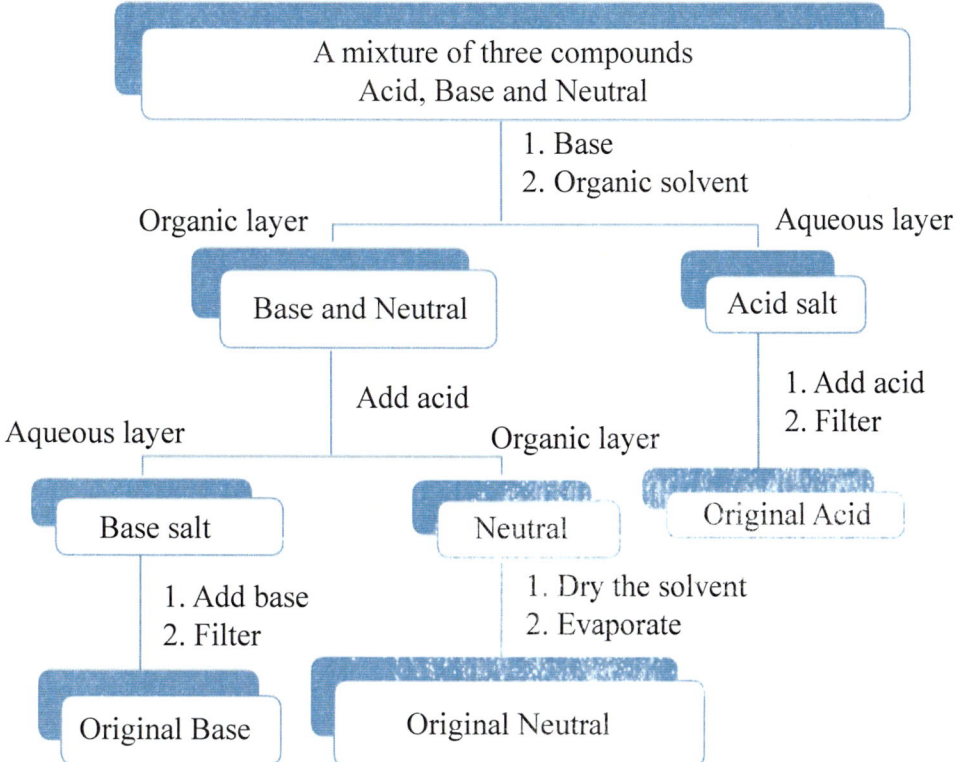

C hapter 7

RECRYSTALLIZATION

When organic compounds are isolated or synthesized in the laboratory, they are rarely pure when first obtained. Therefore, the contaminants must be removed. A very useful method is to recrystallize the solid. A solvent is sought which will (a) dissolve the major component and not the contaminant, (b) dissolve the contaminant but not the major component, or (c) dissolve both but from which one of the components may selectively crystallize on cooling.

Oftentimes combinations of the above are encountered. Ideally, one looks for a solvent in which the major component dissolves when the solvent is hot and crystallizes upon cooling. The technique that is often used is to dissolve the substance in boiling solvent, filter any insolubles by gravitation from the hot solution, and then allow the solution to cool slowly. Rapid cooling causes less perfect crystal formation. An ideal solvent for recrystallization is one that will show a steep solubility/temperature curve for the solute. Water is a poor solvent for recrystallizing sodium chloride since the salt is almost as soluble in water at 25 °C as it is at 100 °C.

In certain instances a single solvent cannot be found for recrystallization of a solute. In those cases it is convenient to dissolve the solute in a "good" solvent, filter hot, and add an anti-solvent at the boiling temperature until the solution becomes cloudy. Then add just enough solvent to clear up the cloudiness.

In order to be more confident of purity, one should recrystallize a substance until a constant and sharp m.p. is obtained. Furthermore, the m.p. of a pure substance will not change when different solvents are used for recrystallization.

The most difficult part in recrystallization is choosing the right solvent. Selecting a solvent based on the structure of the compound is always misleading. Therefore, a good solvent must be chosen based on previous experience or determined experimentally. Here are some characteristics of a good solvent:

1. Dissolve a moderate quantity of solute at high temperatures, and very little at low temperature.

2. Dissolves impurities at low temperature.

3. Does not react chemically with solute.

4. Low toxicity.

5. Low to moderate boiling point.

6. Easily removed from purified substance.

Types of Impurities

1. Traces of coloring matter – removed with decolorizing carbon. Impurities are adsorbed on the surface of carbon (charcoal).

2. Insoluble materials – removed by filtration.

3. More soluble materials than the desired product – easily removed.

4. Less soluble materials than the desired compound – hard to remove.

Procedure

a. Dissolve the solid in a minimum amount of the chosen solvent in an Erlenmeyer flask or Craig Tube. Put just enough solvent to cover the solid.

b. Heat the mixture on a sand bath or hot plate to completely dissolve the compound. Very volatile or toxic solvents should be put in a flask fitted with a reflux condenser while dissolving the solid. If there are impurities, filter the solution by gravity into an Erlenmeyer flask that is heated during the filtration.

c. Allow the solution to cool to room temperature, and then place in an ice-bath for crystals to form.

d. Collect crystals by vacuum filtration using a Hirsch funnel or by centrifugal filtration.

e. Wash the crystals with cold solvent and allow them to air-dry for some time before removing them from the funnel.

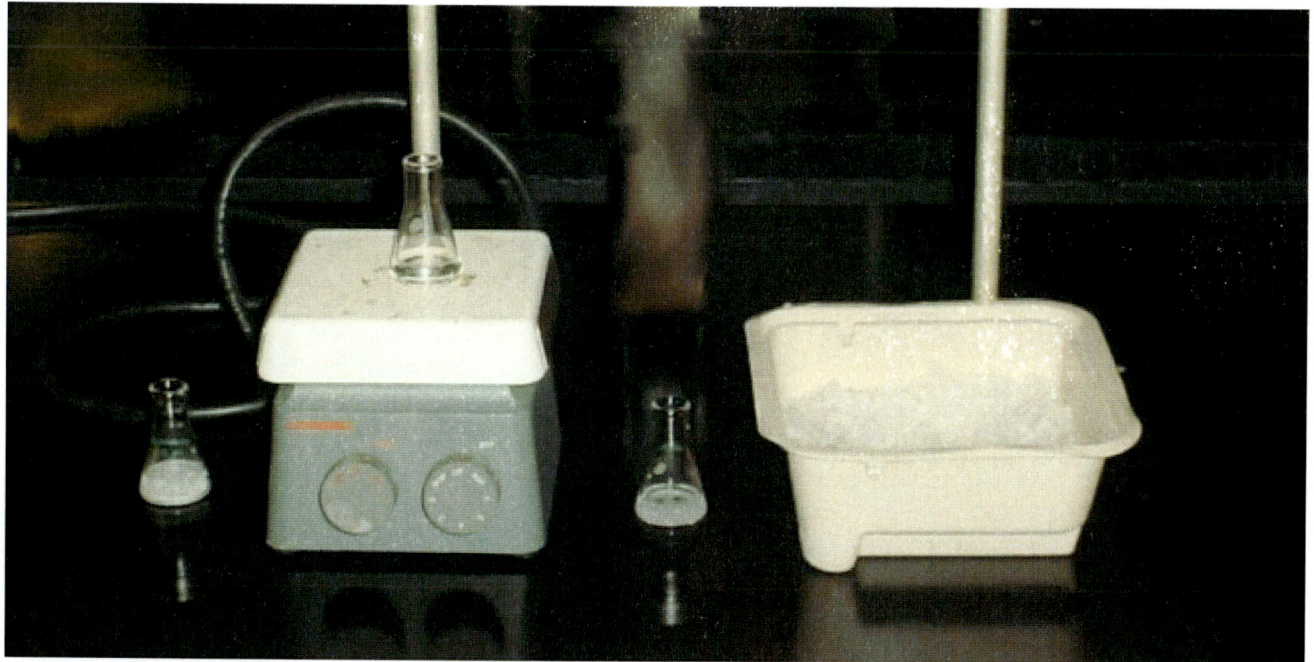

The Process of Recrystallization

The suction flask is connected to the aspirator (fig. A) or vacuum pump (fig. B) as shown below. Put a 400 mL beaker under the aspirator to control the water flowing from it. Using the vacuum pump is more efficient and it saves water.

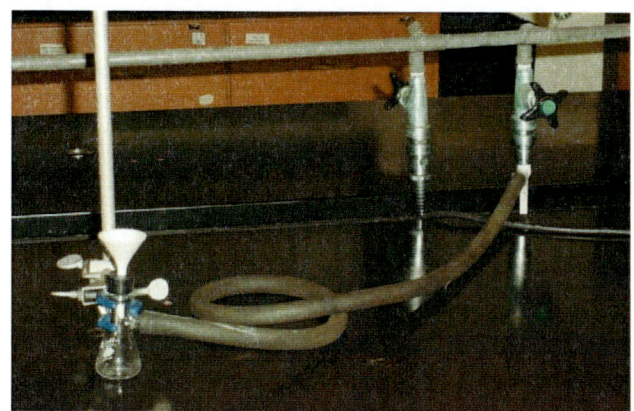

Fig. A

Fig. B

Set–up for Vacuum (suction) Filtration

Centrifugal Filtration

Centrifuge

Chapter 8

CHROMATOGRAPHY

Chromatography is a technique used for analyzing, separating and purifying compounds by distribution between two phases, one of which is stationary, and the other moving. There are different types of chromatography: 1) **Liquid-Solid** (TLC and Column), 2) **Liquid-Liquid** (Paper), and 3) **Gas-Liquid** (Gas or vapor phase). Chromatographic techniques are probably the most universally important of all the skills in which an organic chemist requires expertise.

Thin-layer Chromatography

Thin-layer chromatography is a relatively simple and rapid method for analytical and preparative applications in organic chemistry. It has been used to separate and identify almost all types of organic compounds, including amino acids, sugars, steroids, polymers, as well as more simple substances. In essence, thin-layer chromatography consists of spotting a very small amount (5-50 mg.) of substance on a thin-layer of absorbent. The absorbent is backed by a(n) aluminum, glass, paper or otherwise plate. It is a versatile technique, inasmuch as the absorbents may be many, just as in column chromatography. The most widely used absorbents thus far are silica gel and alumina, but others are useful for certain applications.

The plates are then placed in an enclosed container with just the bottom of the plate immersed in developing solvent. As the solvent moves up the layer of absorbent, it carries the spotted substance at a rate dependent on the type of solvent, the type of substance, and the type of absorbent. If the substance is held loosely by the absorbent it may be taken up the column even by relatively poor solvents. A substance unlike the solvent will travel farther up the column. For example, if a polar solvent is used and a polar substance is spotted next to a nonpolar substance, the nonpolar substance will resist sticking to the plate and thus move farther up the column. The polar substance will stick to the like solvent and resist movement up the column. By choosing the proper absorbent and solvent system, it is possible to separate substances with only minor structural differences. Solvents that are commonly used for TLC are: **Petroleum ether, Diethyl ether, Hexane(s), Ethyl acetate and Acetone.**

In practice, the solvent is allowed to rise to a measured height on the plate (usually near the top) and then the plate removed from the developing chamber. The distance which the spotted substance rises may be determined by observation if the substance is colored or by spraying the plates with reagents which give colored compounds with the separated substances.

Materials for TLC

Procedure

Prepare a TLC plate by taking a 1 inch x 3 inch plate and drawing (with pencil) a straight horizontal line with pencil 1.5 cm above the bottom edge (on the smooth white chalky side). Using the open end of a capillary tube (or a pipet), place a drop of the substance in the center of the line. (If using a capillary tube, spot the plate twice in the same spot.) Draw another horizontal line ~1.5 cm below the top edge.

Prepare a developing tank by fitting a piece of filter paper in a 250 ml. Beaker (or a cylinder at least 12 inches high if the larger plates are used) and adding enough solvent so that the level of solvent is 0.5 cm and the filter paper is wet with solvent. Cover the tank with an inverted watch glass. When equilibrium (about 5 min) has been established in the tank place the plate in the developing chamber. When the solvent has reached the top line, remove the plate and allow it to dry. Observe the spots under an ultraviolet (UV) lamp and circle the spots with a pencil. Calculate the R_f values for each of the spots.

Experiment in Progress

Each substance is assigned an R_f value in a particular system by dividing the distance moved by the substance by the distance moved by the solvent. For example, if substance A moves 4.0 cm On a plate while the solvent moves 8.0 cm, the R_f value for A is: $\dfrac{4.0}{8.0}$ or 0.5.

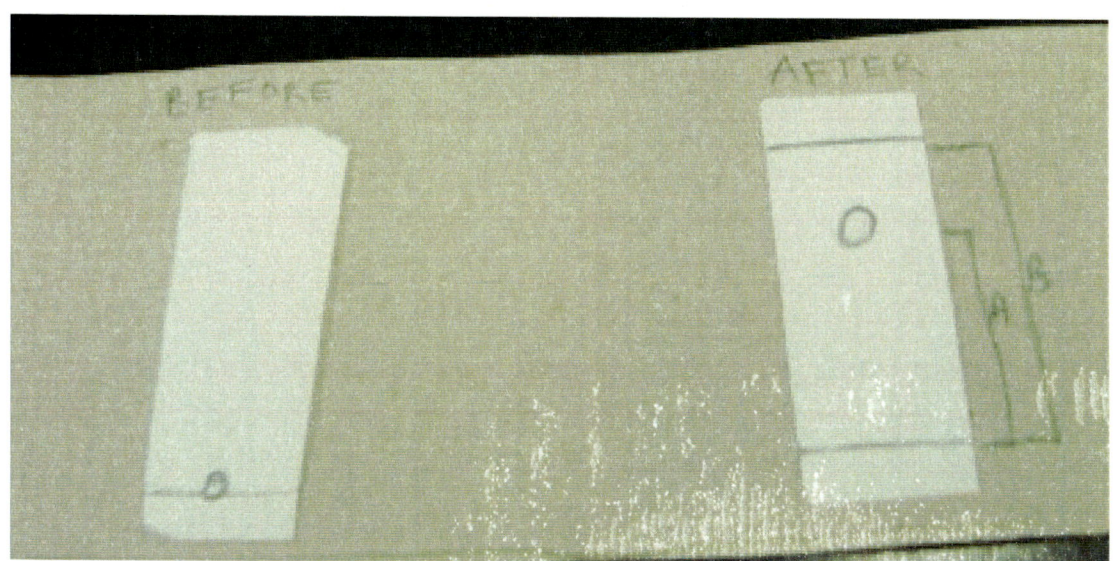

At The End of The Experiment

Paper Chromatography

Paper chromatography is a typical partition system, in which the paper serves as the support for the stationary phase, adsorbed water. The paper is almost pure cellulose and in a water-saturated atmosphere, adsorbs more than 20% water. The mobile phase is usually a mixture of organic solvents and water. The mobile phase travels up the paper (ascending chromatography) from a reservoir or down the paper (descending chromatography) or radially and horizontally from a central spot on a circular piece of paper.

The mixture to be separated is spotted on the paper near the reservoir of the mobile phase. As the mobile phase moves by, the mixture is taken with it at rates depending on the particular components, the solvents used, type of paper, temperature, and other factors. Ideally, each component of the mixture will move at a different rate on the paper. After the solvent front has moved a certain distance on the paper, the paper is removed from the chromatographic apparatus and the distance moved by various components of the mixture determined. If the substances are colored, this is easily accomplished. If the substances are not colored, some reagent (Ninhydrin) must be sprayed on the paper to produce a color with the components.

The relative distances traveled by the components are expressed as R_f values, where

$$R_f = \frac{distance\ traveled\ by\ the\ component}{distance\ traveled\ by\ the\ solvent\ front}$$

In paper chromatography, the stationary phase is quite hydrophilic. Therefore, it would be expected and has been experimentally verified that the more hydrophilic a component of a mixture is, the less distance it will be carried by the mobile phase. Paper chromatography requires only minimum amounts of materials. It has found wide applicability especially in the separation and identification of amino acids and sugars. Some factors which affect the solubility of amino acids in water are: 1) Polarity—As more polar

groups are added to the molecule, water solubility increases, and the R_f value decreases. 2) Molecular weight—As the molecular weight increases, water solubility decreases, and the R_f value increases.

Technique

1. Do not touch the paper with fingers. Contact of paper with skin protein usually gives false results (additional spots).

2. Procedure is similar to that of thin layer chromatography (TLC).

3. Fold about 1 cm of the top end of the paper at right angles and staple it. Repeat with the lower end.

4. Lower the paper into the container (developing chamber) so that it does not touch the sides.

5. Allow the chromatogram to develop until the solvent reaches the finish line 1-2 cm from the top.

6. Remove the paper from the container and suspend it in air to dry.

7. To detect the amino acids, spray the paper with a 2% solution of ninhydrin in ethanol.

8. Calculate the R_f value.

Gas Chromatography

In gas chromatography, the partitioning processes for the mixture to be separated are carried out between a stationary liquid phase and a moving gas phase. In the typical operation of gas chromatography, the sample is injected into **Port B** and is immediately vaporized and introduced into a moving stream of gas, called the carrier gas. The sample then passes through a column (contained in a temperature-controlled oven) filled with particles coated with liquid adsorbent. During this period, the sample is exposed to many gas-liquid partitioning processes and is separated. Each component exits the column and its presence is detected by an electrical detector that generates a signal, which is recorded on a chart recorder.

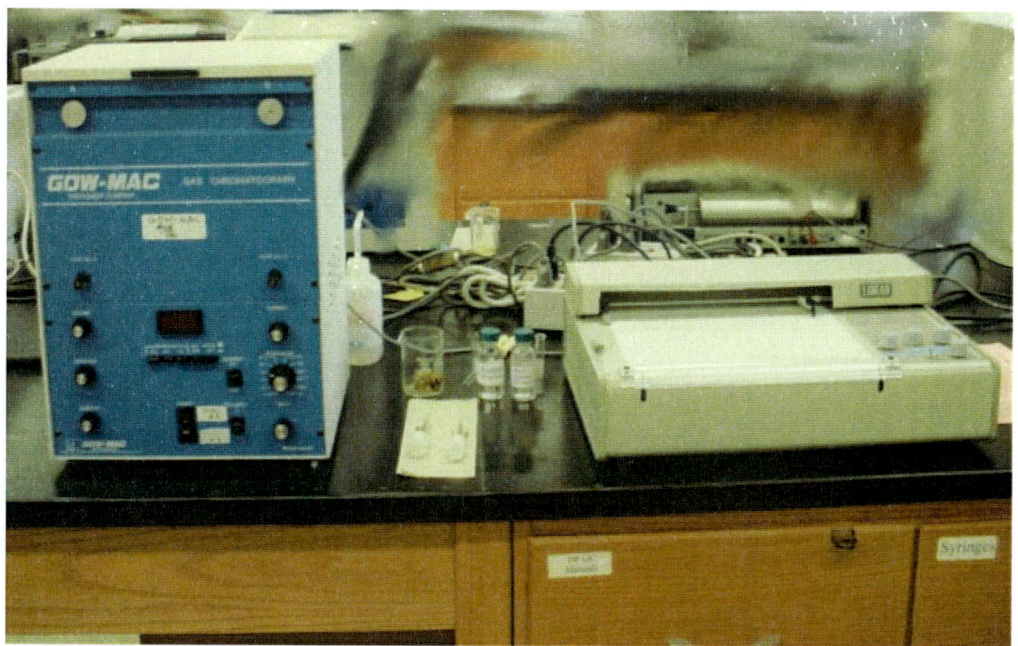

GOW-MAC Manual Gas Chromatogr

Procedure for Operating the GOW-MAC Gas Chromatograph

1. Make sure the temperature of the chromatograph is set at less than 20 °C below the boiling point of your substance.

2. Rinse the syringe with acetone twice then several times with your sample.

3. Fill the syringe with your sample.

4. Inject the sample into Port B of the GC.

5. Place a mark on the GC paper immediately after injecting the sample to indicate the beginning of the process.

6. Stop the recorder after all the peaks have appeared and you observe a straight line for a minute.

7. Clean the syringe with acetone when you are done with all your samples.

Factors Affecting Separation

1. Column (length or packing). This is the most important part of the gas chromatograph.

2. The rate of flow of the carrier gas.

3. Column Temperature: 1) Too high—entire mixture is flushed through the column at the same rate as carrier gas. The result is no separation. 2) Too Low—the mixture dissolves in the liquid phase and never vaporizes. No separation is observed.

4. Polar samples stick to the polar column

5. Non-polar samples stick to the non-polar column

6. Boiling point also affects the movement of samples. Low boiling point samples always come out first.

Comparison of GC and TLC

	TLC	GC
Mobile Phase	Solvent	Carrier gas (He, N_2, Ar) Gas is inert (no interaction with sample)
Stationary Phase	Silica gel	Column packing (liquid polymer)

Automatic Gas/Mass Chromatography

Procedure for GCMS

1.   Place liquid into a vial (fill it halfway or more) and put a lid on the vial.

2.   Place the vial into an open slot and remember the number of the slot it is placed in. This is the vial number.

3.   Open up **GCMS_1.** on the computer.

4.   Go to **Methods.** Click **Load Method.**

5.   Select Folder **CHEM233_FA13** from the list of folders.

6.   Select **SIMPLE_DISTILLATION_2.m** from the list of methods.

7.   Go to **Methods.** Click on **Run Method.**

8.   Input the vial number where it is asked for in the **Run Method** screen.

9.   Fill in a file name in order to find your graph later.

10. Click **Ok and Run Method.**

11. Wait for run to end.

12. Open **GCMS_1 Data Analysis** from the desktop.

13. Select **File**. Click on **Load File**.

14. Select your file name from the list of file names that pop up.

15. The GC graph will appear in the Data Analysis screen, double right click on the peak for the MS to appear.

16. Select **Print**. Make sure **TIC and Spectrum** is chosen.Click **Ok** to print.

Chromatogram

This is the printout you get after performing gas chromatography. Two important pieces of information come from the chromatogram:

Sample GC/MS Chromatogram

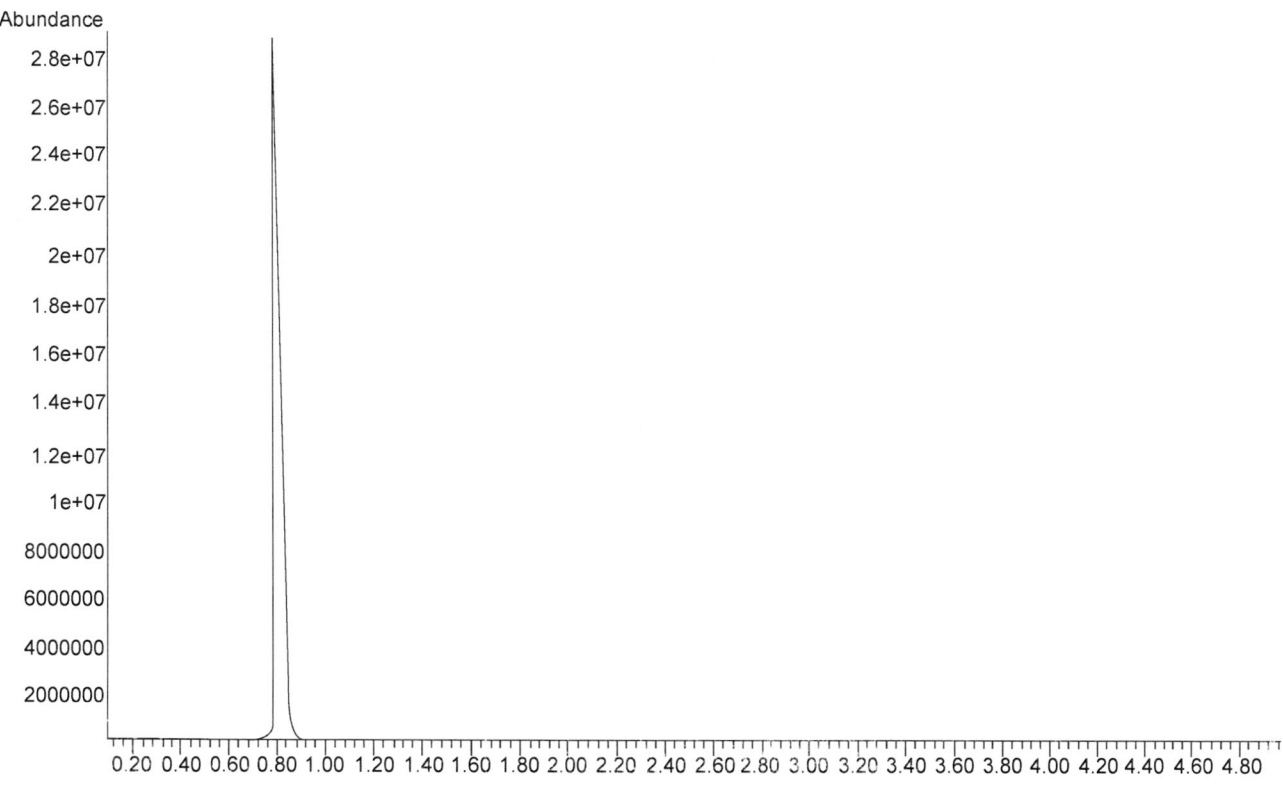

TIC:METHANOL_STD_4.D\data.ms

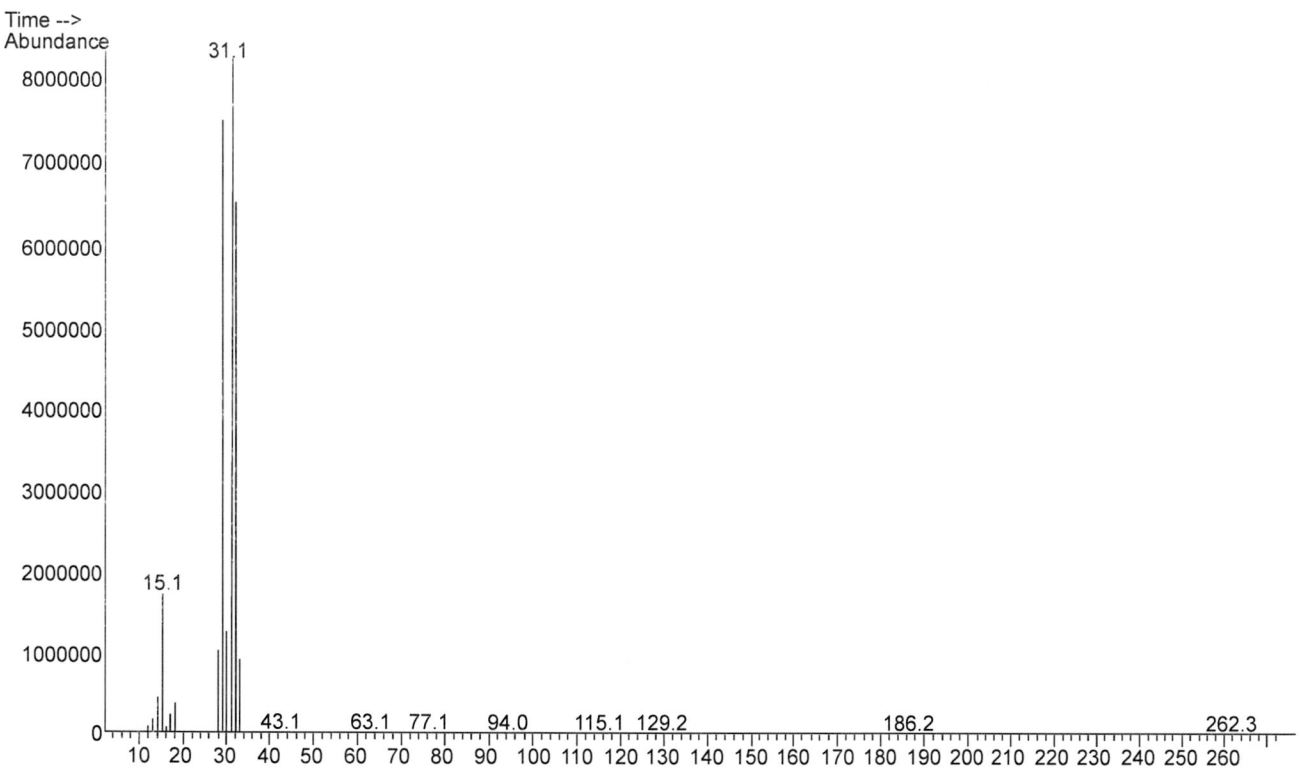

Scan 60 (0.805 min): METGANOL_STD_4.D\data.ms

1. Retention Time (R_t)

This is the time it takes for a compound to pass through the column after injection. This is a measure from the time of injection to the time of maximum deflection of the pen for each component.

$$R_t = \text{Distance (cm)} \times 1/\text{chart speed (cm/min)}$$

2. Proportion of Samples

The area under a peak is proportional to the amount of that component in the mixture. There are different methods that are used to determine the area under a gas chromatograph peak. We are going to use the Triangulation of a Peak method.

Triangulation Method

•   Measure the height of the peak

•   Measure the width at ½ height

•   Area of peak = height × width at ½ height

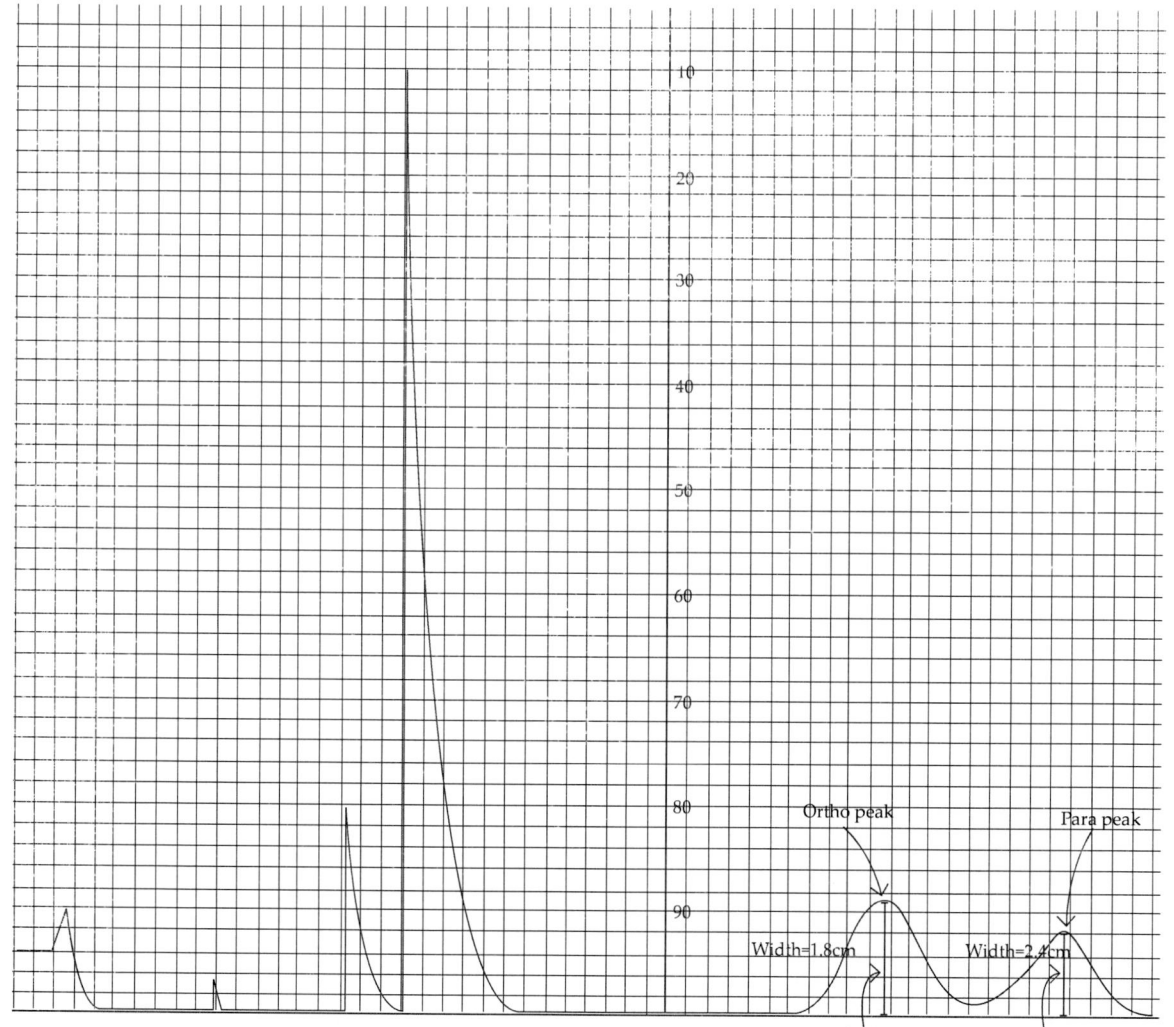

Sample GC Chromatogram

To determine the distribution of isomers or component in a mixture

Using triangulation method, the percentage of each isomer or component in a mixture can be calculated.

% Peak A = $\dfrac{\text{Area peak A}}{\text{Total area of all peaks}} \times 100$

% Peak B = $\dfrac{\text{Area peak B}}{\text{Total area of all peaks}} \times 100$

% Peak C = $\dfrac{\text{Area peak C}}{\text{Total area of all peaks}} \times 100$

SPECTROSCOPY

Many organic chemical identification problems have been vastly simplified by the use of spectrographic methods. Advances in instrumentation, especially since World War II, have not only shortened the time for identification of organic compounds but have made it possible to identify some otherwise unidentifiable compounds.

Although the field of spectroscopy includes both emission and absorption spectroscopy, we will be concerned only with the latter. Absorption spectra are obtained by placing samples between the spectrometer and a source of electromagnetic radiation having a given frequency range. The spectrometer analyzes the energy transmitted relative to the incident energy. Most of the absorbed energy in the infrared region produces heat, while that absorbed in the ultraviolet region is re-emitted as light. Ultraviolet spectroscopy will be briefly mentioned. It is primarily useful with conjugated systems.

The mechanisms of energy absorption in the various frequency regions vary considerably but in all cases a <u>certain</u> amount of energy is absorbed – that amount required for a transition from a lower to higher energy state. The energy absorbed anywhere in the electromagnetic spectrum is described by:

$$\Delta E = hv$$

$$\Delta E = \text{change in energy}$$

$$h = \text{Planck's constant}; 6.624 \times 10^{-27} \text{ erg-sec.}$$

$$v = \text{Frequency of incident light (cycles/sec; cps).}$$

An Overview of Spectroscopic Methods

IR Spectrum

The infrared spectrophotometer records the percent transmittance of incident light through the sample as a function of its wavelength. The presence or absence of major functional groups is detected because the absorption band of a functional group is characteristic of that group regardless of the other structural features of the molecule. Its position lies within an identifying range. Subtleties in placement are due to electronic effects, resonance, and hydrogen bonding.

¹HNMR Spectrum

The number, position, intensity, and splitting pattern of signals in the nuclear magnetic spectrum provide information about the symmetry, electronic environment of the proton, quantity of protons present, and nature of adjacent protons in the molecule. The graph plots the chemical shift in ppm (parts per million) versus the intensity as a percentage of the base peak.

UV Spectrum

The ultraviolet (UV) spectrum reveals the degree of unsaturation and quantitatively measures the extent of conjugation of the compound. The spectrum describes the position of the maximum absorptions on the x-axis and the intensity of the absorbance of the incident light on the y-axis.

MS Spectrum

This spectrum is primarily used for the determination of the molecular weight of a compound, which is numerically equivalent to the mass to charge ratio of the molecular ion. The spectrum is a plot in which the x-axis represents the mass to charge ratio, and the y-axis represents the relative number of ions.

The discussion here will be confined to infrared and nuclear magnetic resonance spectroscopy for the most part since these are most useful to the organic chemist.

A. Infrared Spectroscopy

The infrared spectrum of an organic compound provides a wealth of information about its structure. In the infrared region it is customary to express wavelength in microns, u (u = 10^{-4} cm or 10^{40}A), and frequency in wave numbers or reciprocal centimeters, cm^{-1}. The infrared region extends from 4000 to 500 cm^{-1}. There are two regions in an IR spectrum: 1) **Primary region**, functional group region (4000 to 1500 cm^{-1}) and 2) **Finger Print region**, confirming evidence region (1500 to 500 cm^{-1}). The wave number is directly proportional to absorbed energy (k = $\Delta E/hc$) whereas wavelength is inversely proportional to the energy absorbed.

Molecules are not rigid but are constantly vibrating. They may be pictured as composed of balls connected together by springs. Molecules undergo two main types of vibrations: **stretching**, in which the distance between two atoms increases or decreases while the bond angles do not change; and **bending**, in which bond angles do change.

The great value of infrared spectroscopy to the organic chemist lies in the fact that each group absorbs in a rather narrow range. For example, the O-H stretching frequency is found in the 3200-3600 cm^{-1} region; the greater the hydrogen bonding the lower the frequency. The C=O group absorbs from about 1680 to 1760 cm^{-1} depending on whether it is an aldehyde, ketone, ester, or acid group. A rather complete listing of functional groups and their absorption bands may be found at the end of this Section. The infrared spectrum, therefore, tells us what groups are present (or absent) in a molecule.

Interpretation of an infrared spectrum is not, however, a simple matter. In addition to the fundamental bands there are overtones (harmonics) that may appear at just twice the frequency of the fundamental band. Furthermore, absorption bands of certain groups may be shifted because of conjugation, electron withdrawal or donation by a neighboring group, angle strain, or by hydrogen bonding. It has already been mentioned that hydrogen bonding causes a shift to a lower frequency for the O-H stretching vibration. Hydrogen bonding weakens the O-H band and therefore less energy is required to stretch it.

Knowledge of structural organic chemistry is of great help in predicting shifts in absorption. For example, it is believed that in, β-unsaturated aldehydes and ketones the double bond character is decreased in both the carbonyl and C=C groups. It is not surprising then to find that unsaturated ketones show carbonyl (C=O) absorption at about 1685 cm^{-1}. (a decrease of about 25 cm^{-1}).

Chromophores

- The bands that are absorbed correspond to an identifiable part of the molecule known as a chromophore.

- Chromophores often correspond to very specific functional groups.

- By determining the bands absorbed, one can determine the functional groups present in a molecule.

- No two compounds give the same IR spectrum except for enantiomers.

- It is NOT possible to use IR only to determine the structure of a compound. You can use it to determine functional groups present in a compound.

- Intensity of band is propotional to change in dipole moment that occurs on stretching the bonds.

- Polar bonds give strong IR absorptions.

- Symmetrical bonds may not absorb at all.

Factors that Determine IR absorption Position (cm^{-1})

1. Bond strength

2. Masses of atoms involved in bond

3. Type of vibration (stretch or bend)

Important Chromophores

- Hydroxyl (O-H) stretch

- Carbonyl group

- Polar functional groups (N-H, etc...)

- Chromophores absorbing in the region between 1900 and 2600 cm-1

  Acetylenes

  Nitriles

  Allenes

  Diazo compounds

  Isocyanates

  Isothiocyanates

Three Different Types of Spectrophotometers are presented

1.   Perkin Elmer Spectrum One FTIR

Spectrophotometer, Monitor, CPU, Printer

Spectrophotometer and Monitor

Procedure

Infrared spectra may be determined on liquids, solids, and gases. In this course, we are going to deal only with liquids and solids.

- Place sample (solid or liquid) in the small well on the spectrophotometer.

- Use the micropress to press on the solid by lowering the micropress as far down as it will comfortably go.

- Use the mouse to select Scan.

- Wait for the spectrum to appear on the screen.

- Select ACCEPT if an error message appears.

- Name your spectrum.

- Select OPTIMIZE from the VIEW MENU.

- Select LABEL PEAKS from the VIEW MENU.

- Use the mouse and select PRINT from the screen.

- Clean the well with acetone for the next sample.

2.   Agilent Technologies Cary 630 FTIR

Procedure

1.   Open METHODS. Choose the Method Pathlength_Transmission.

2.   Go back to the HOME screen.

3.   Select START. Follow the instructions on the screen.

   a.   Clean IR sample plate.

   b.   Click NEXT, wait while it performs the background scan.

   c.   Place solid or 1 drop of sample on the sample plate. For solid, lower the press until you hear a sound.

   d.   Click NEXT and wait until Sampling is over.

4.   Once the IR is done scanning click the DETAILS button.

5.   If whole sample is not shown, click the SAMPLE button.

6.   Right click on the graph, select PICK PEAKS.

7.   Draw a line above the peaks you want to label.

8.   Press BACK.

9.   Select DATA HANDLING.

10.   Select REPORT. Print out the report.

3.   Jasco FT/IR-4600

Procedure

1.   Turn-on the instrument.

2.   Open spectra manager

3.   Double click spectra measurement

4.   Click measure, Start

5.   Select start for pop up window

6.   Select background, wait for sound to stop

7.   When it says "ready" place sample on the crystal and use the micro press for solid

8.   Press start on the IR instrument

9.   Switch to spectra analysis window on bottom of screen

10.   Select processing, peak processing, peak find

11.   Click on desired peaks that aren't already labeled (indicated by red line). Click "add".

12.   Select unwanted peaks and click delete

13.   Click OK when finished

14.   Select file - Print

15.   Clean crystal plate.

TABLE 1

INFRARED ABSORPTION BANDS

A Simplified Correlation Chart

	Type of Vibration		Frequency (cm^{-1})	Intensity
C-H	Alkanes	(stretch)	3000-2850	s
	-CH$_3$	(bend)	1450 and 1375	m
	-CH$_2$-	(bend)	1465	m
	Alkenes	(stretch)	3100-3000	m
		(out-of-plane bend)	1000-650	s
	Aromatics	(stretch)	3150-3050	s
		(out-of-plane bend)	900-690	s
	Alkyne	(stretch)	ca.3300	s
	Aldehyde		2900-2800	w
			2800-2700	w
C-C	Alkane		Not interpretatively useful	
C=C	Alkene		1680-1600	m-w
	Aromatic		1600 and 1475	m-w
C≡C	Alkyne		2250-2100	m-w
C=O	Aldehyde		1740-1720	s
	Ketone		1725-1705	s
	Carboxylic Acid		1725-1700	s
	Ester		1750-1730	s
	Amide		1680-1630	s
	Anhydride		1810 and 1760	s
	Acid Chloride		1800z	s
C-O	Alcohols,Ethers,Esters,Carboxylic Acids, Anhydrides		1300-1000	s
O-H	Alcohols, Phenols			
	Free		3650-3600	m
	H-bonded		3400-3200	m
	Carboxylic Acids		3400-2400	m
N-H	Primary and secondary amines and amides			
	(stretch)		3500-3100	m
	(bend)		1640-1550	m-s
C-N	Amines		1350-1000	m-s
C=N	Imines and Oximes		1690-1640	w-s
C≡N	Nitriles		2260-2240	m
X=C=Y	Allenes, Ketenes, Isocyanates, Isothiocyanates		2270-1940	m-s
N=O	Nitro (R-NO$_2$)		1550 and 1350	s
S-H	Mercaptans		2550	w
S=O	Sulfoxides		1050	s
	Sulfones,Sulfonyl Chlorides, Sulfates, Sulfonamides		1375-1300 and 1350 1140	s
C-X	Fluoride		1400-1000	s
	Chloride		785-540	s
	Bromide,Iodide		<657	s

INFRARED ABSORPTION BANDS

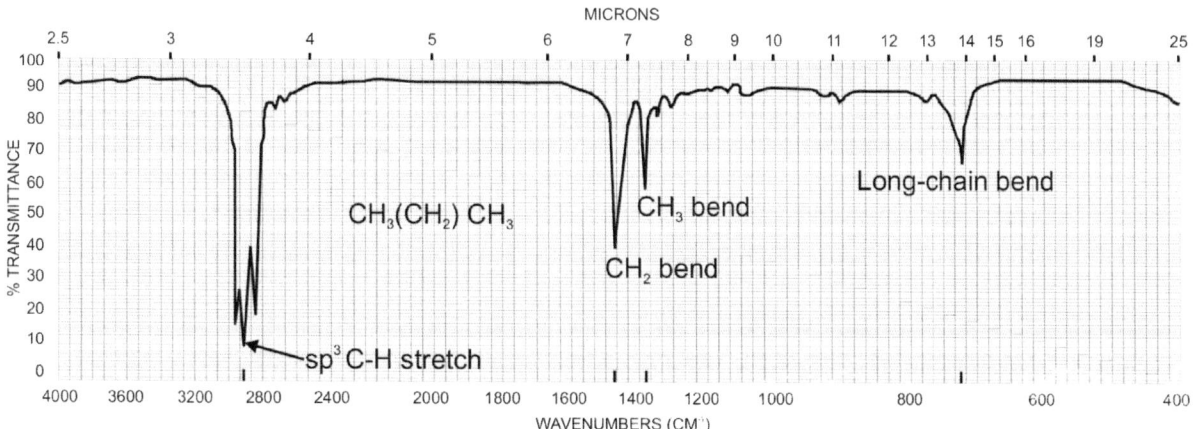

The infrared spectrum of decane

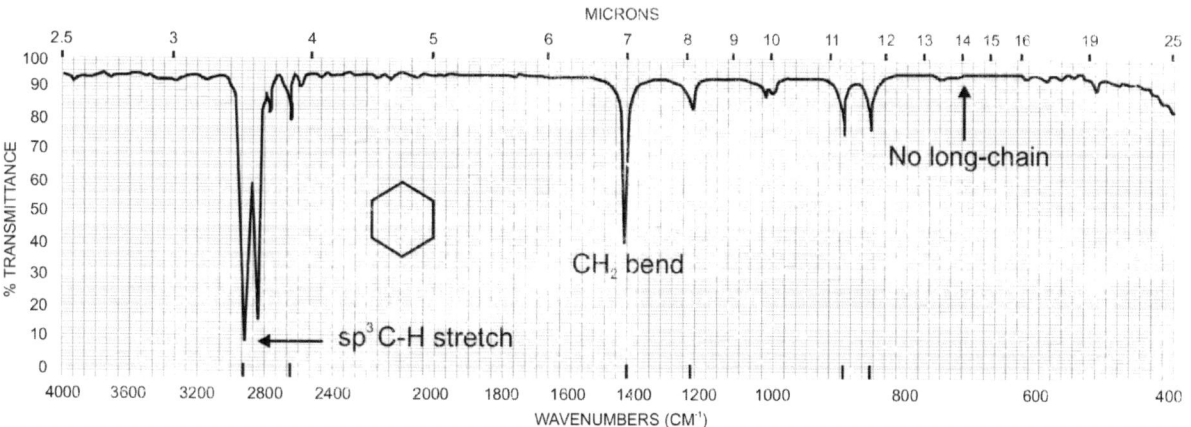

The infrared spectrum of cyclohexane

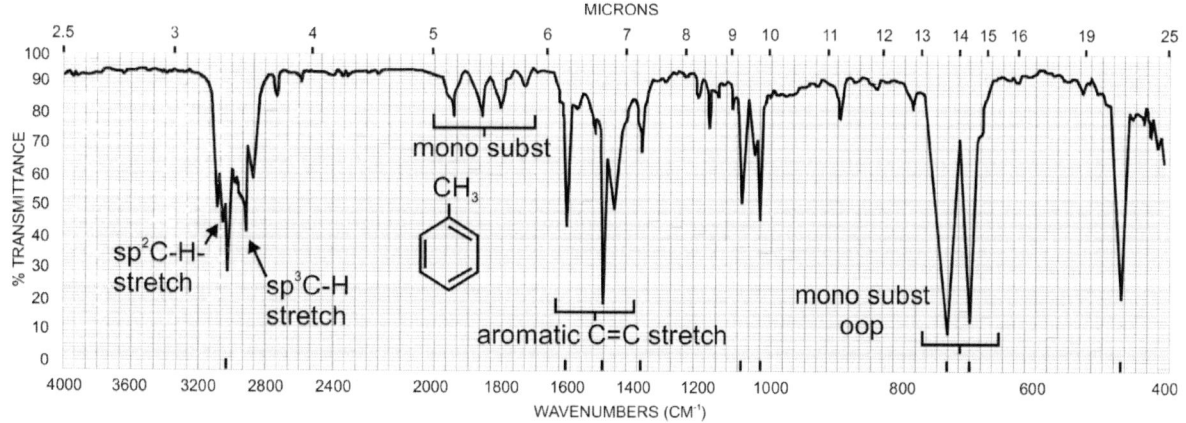

The infrared spectrum of toluene

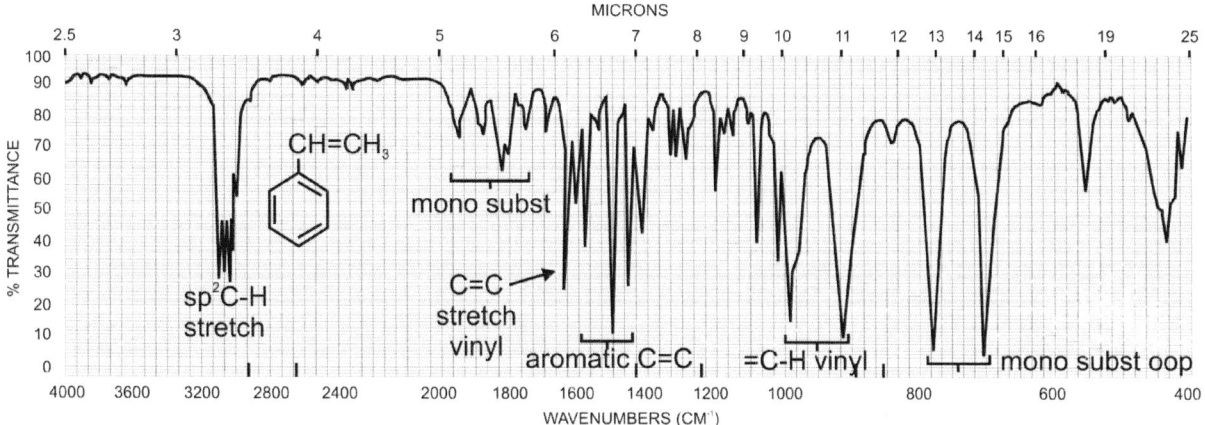

The infrared spectrum of styrene

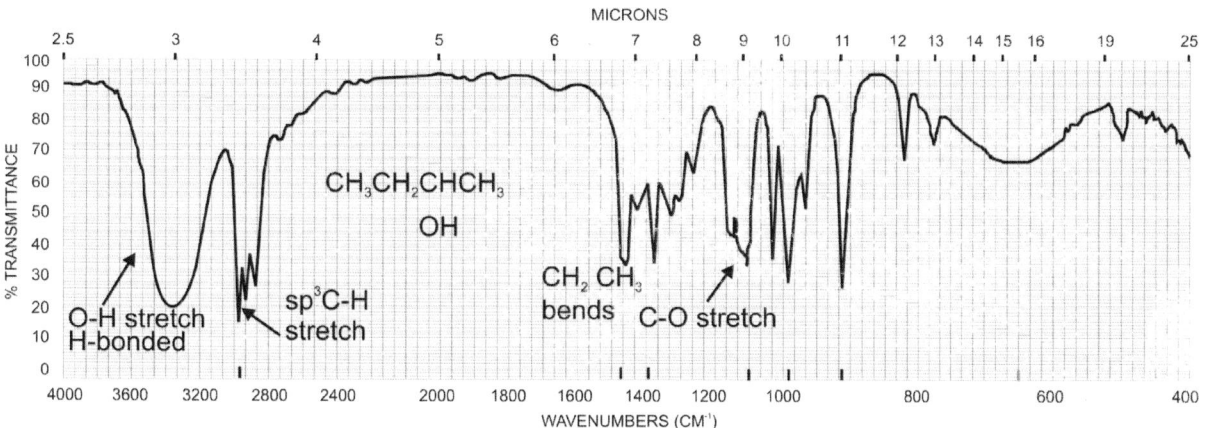

The infrared spectrum of 2-butanol

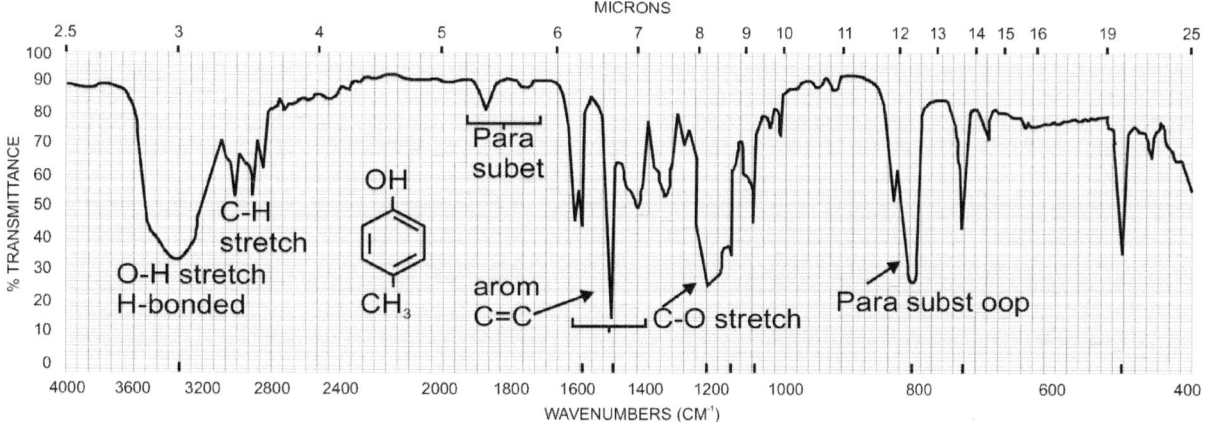

The infrared spectrum of papa-cresol

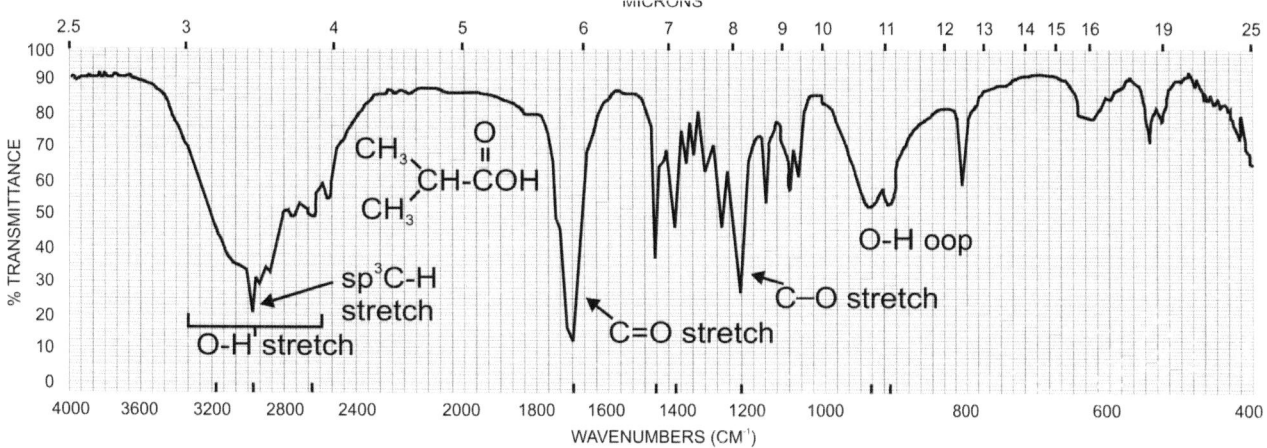

The infrared spectrum of isobutyric acid

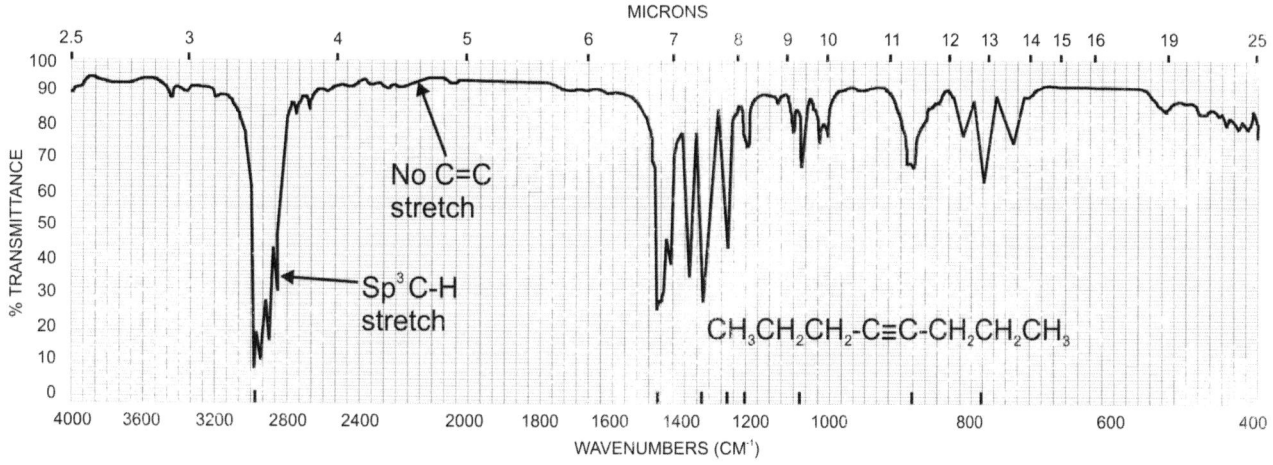

The infrared spectrum of 1-octyne

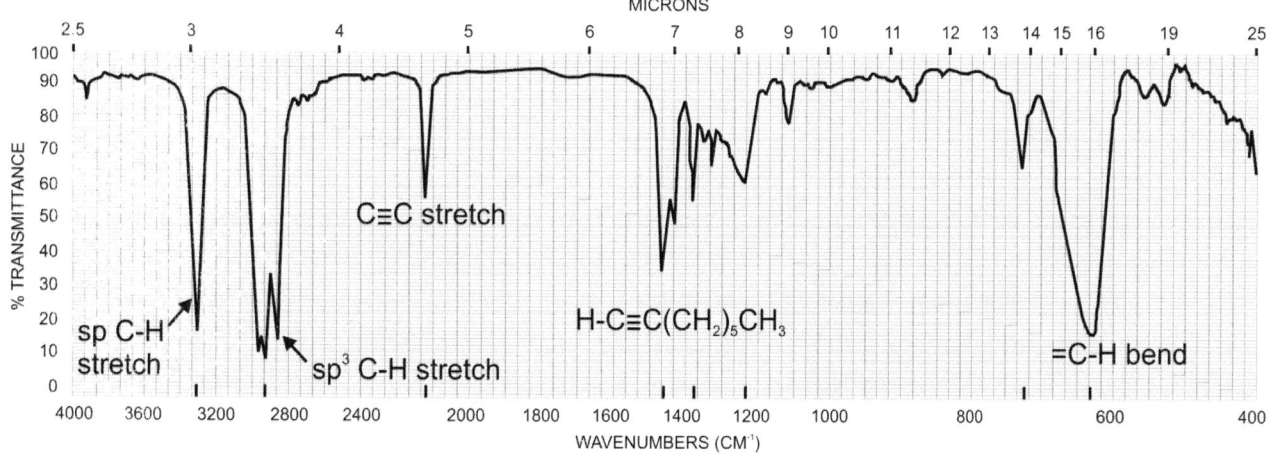

The infrared spectrum of 4-octyne (neat liquid, KBr plates)

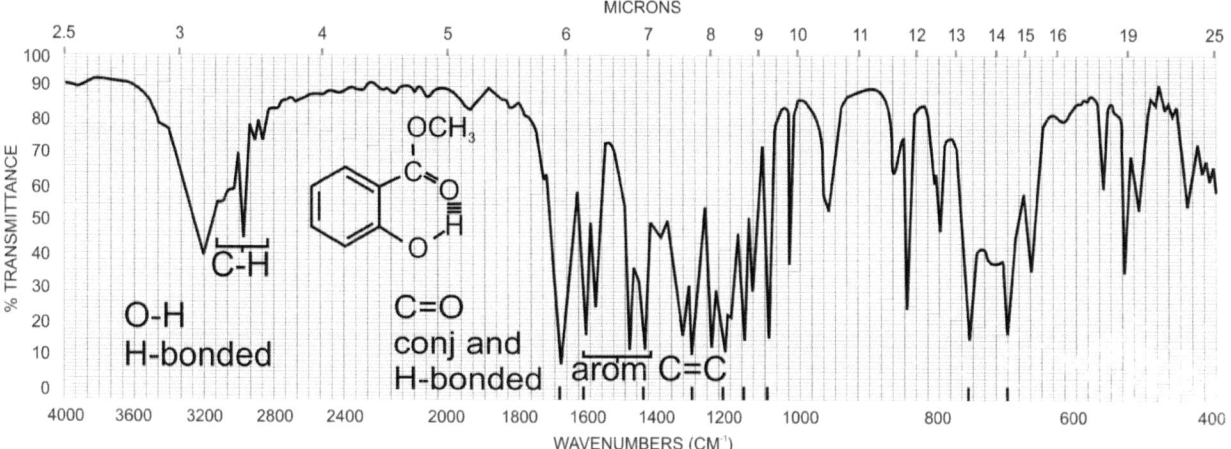

The infrared spectrum of crotonaldehyde

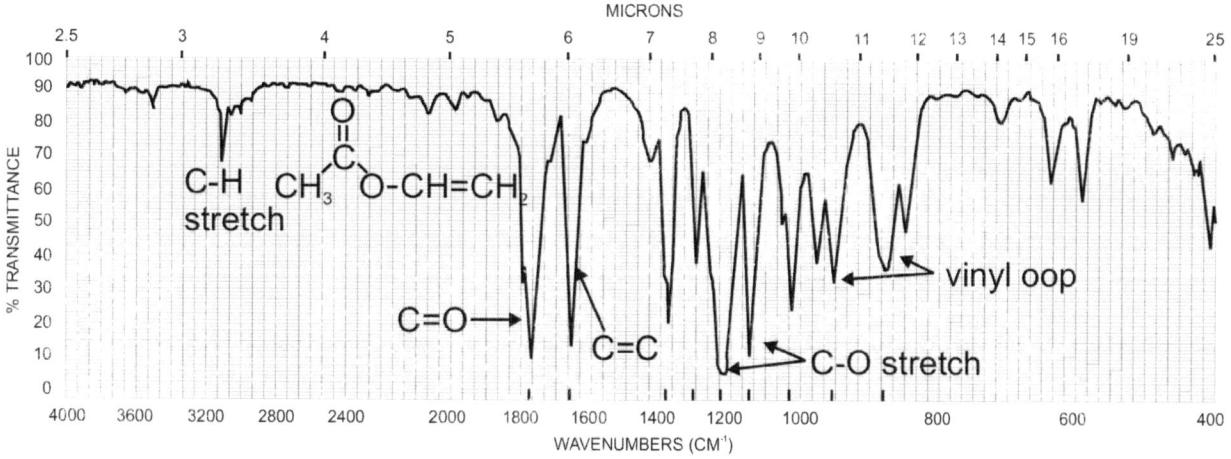

The infrared spectrum of vinyl acetate

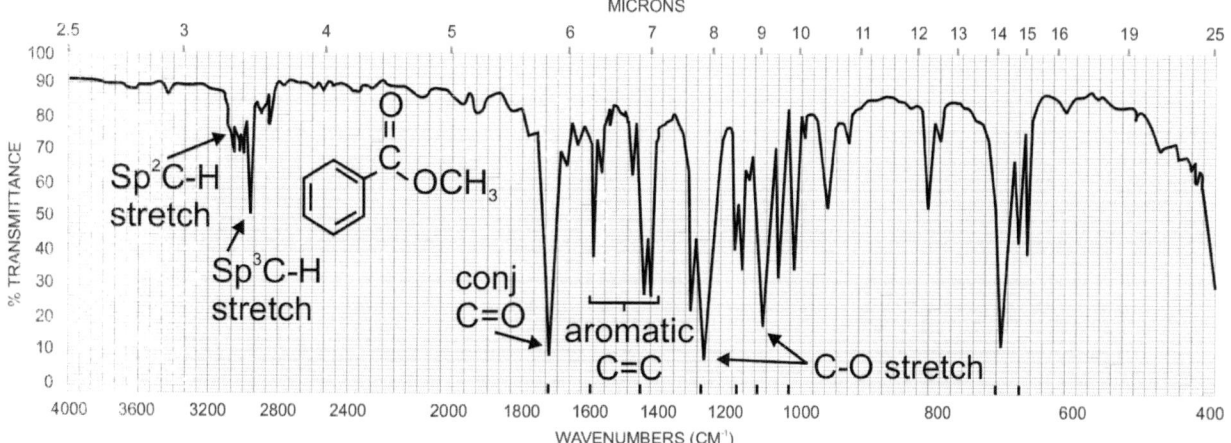

The infrared spectrum of methyl benzoate

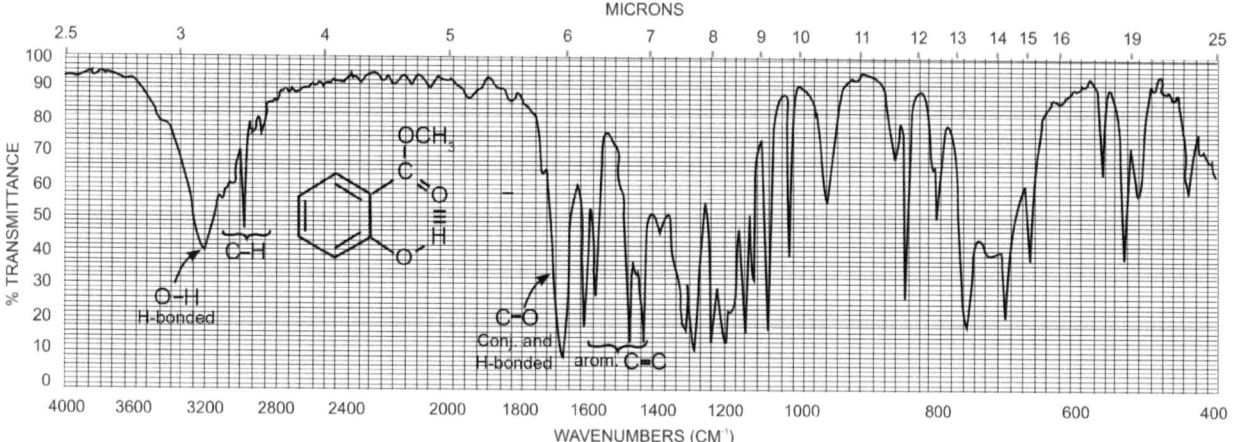

The infrared spectrum of methyl salicylate

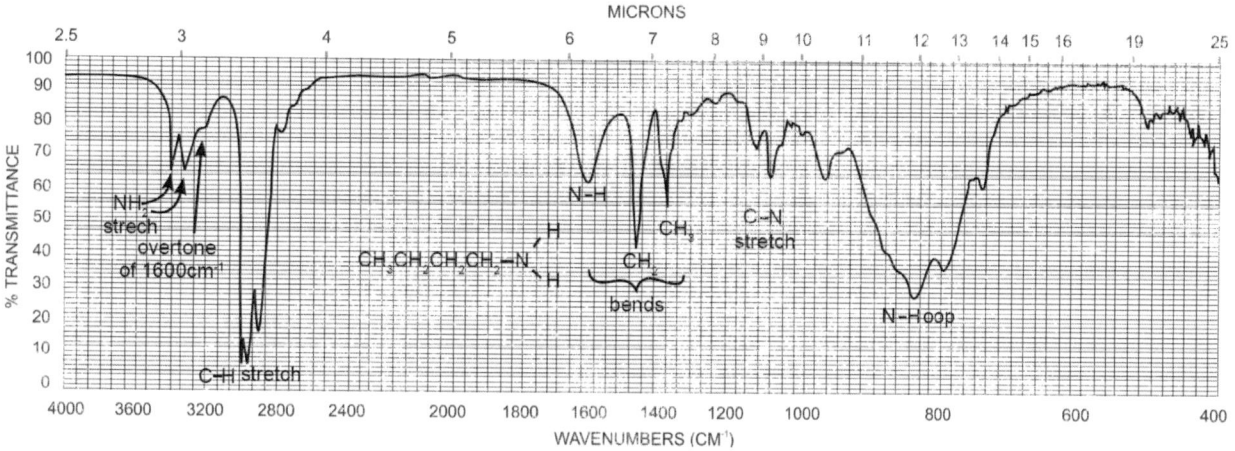

The infrared spectrum of butylamine

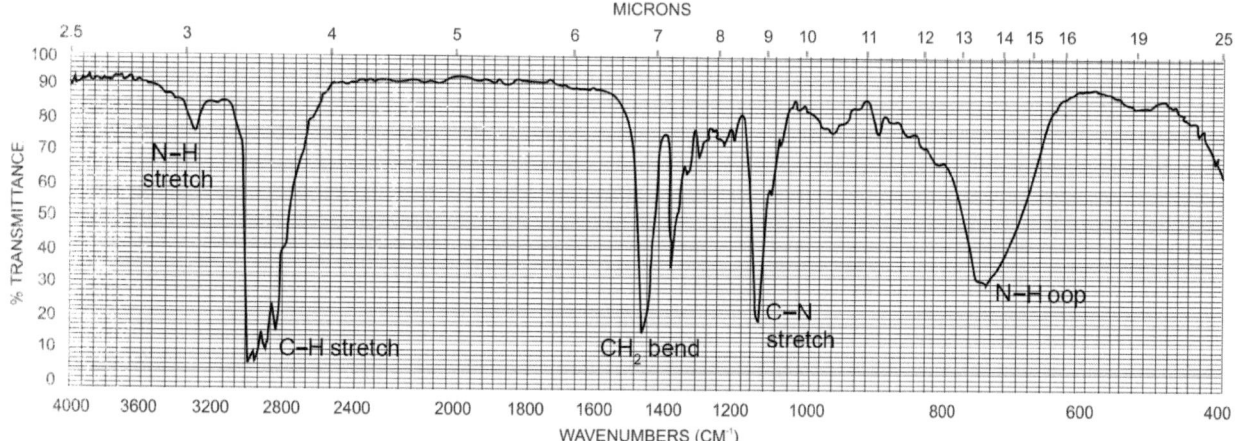

The infrared spectrum of dibutylamine

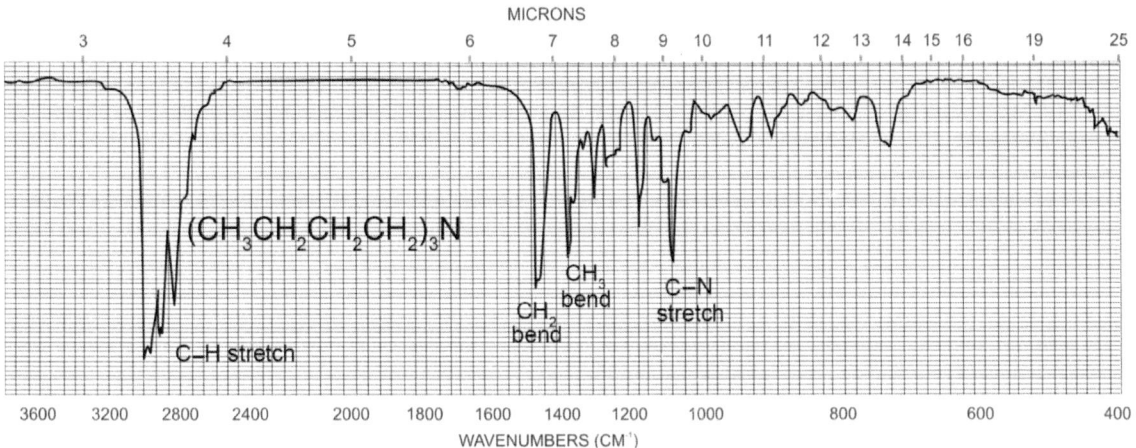

The infrared spectrum of tributylamine

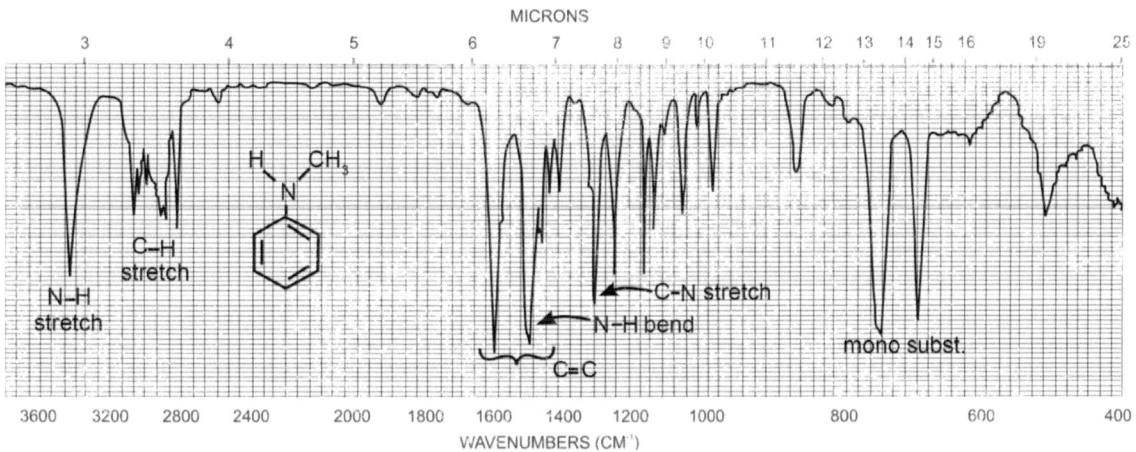

The infrared spectrum of N-methyaniline

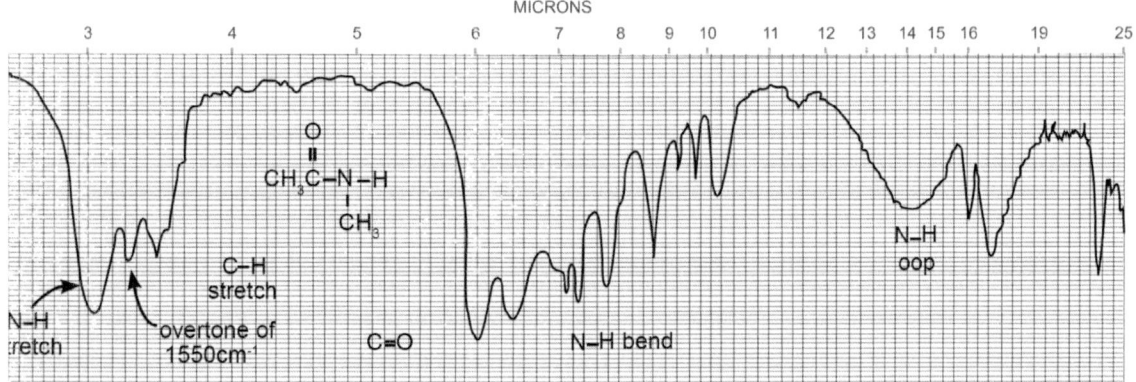

The infrared spectrum of N-methylacetamide

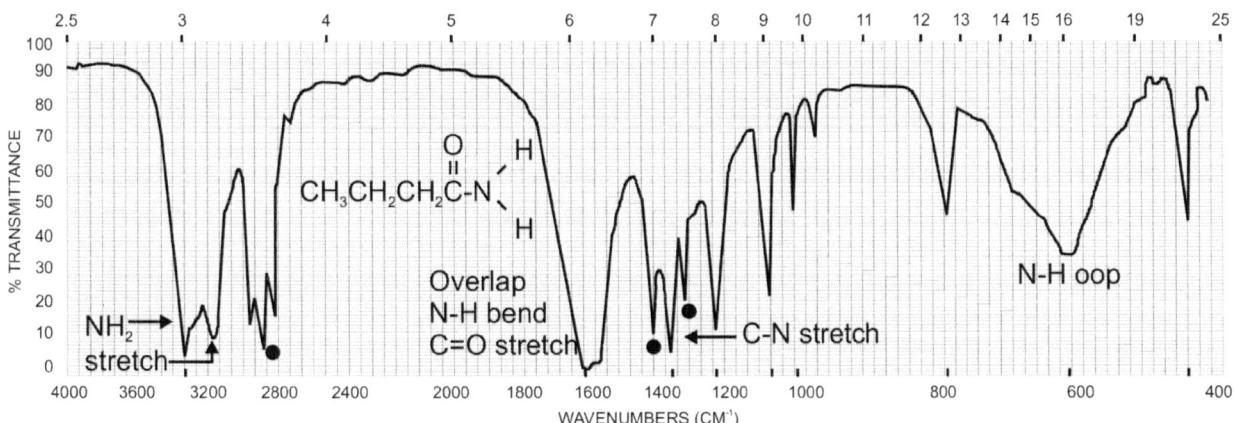

The infrared spectrum of propionamide

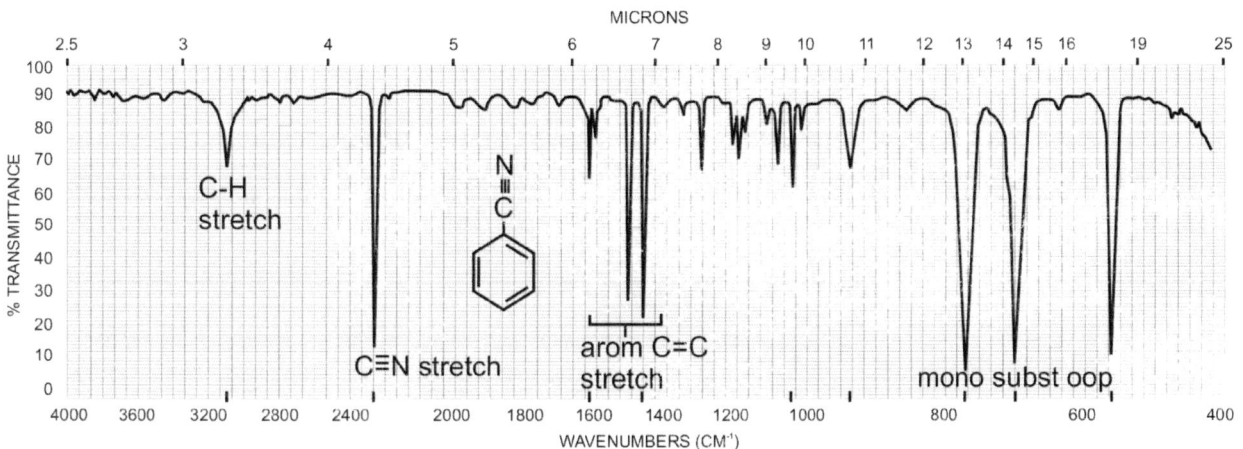

The infrared spectrum of benzonitrile

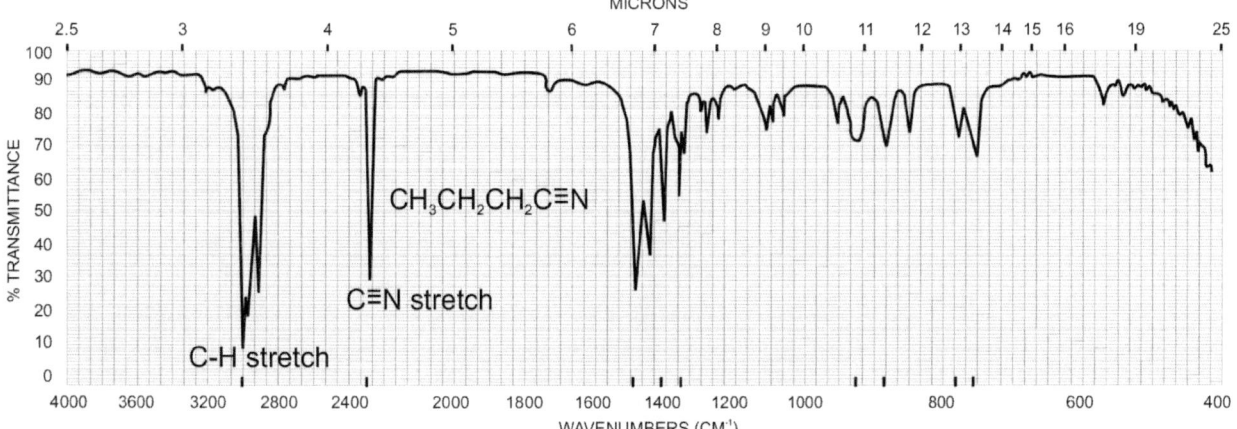

The infrared spectrum of butyronitrile

¹HNMR Spectroscopy

The nuclei of certain atoms possess a mechanical spin or angular momentum. More specifically, magnetic properties are found with nuclei of odd-numbered masses (^{1}H, ^{13}C, ^{17}O, ^{19}F, ^{31}P, etc.) and nuclei of even masses but odd atomic numbers (^{2}H, ^{10}B, ^{14}N, etc.). Such nuclei act like tiny bar magnets and when placed in a magnetic field line up with or against the external field. If the two possible spin states in a collection of nuclei were populated equally the probability of absorbing energy to go to the higher energy state (non-alignment with the field) would be the same as the probability of emitting energy to go to the lower energy state. Consequently, there would be no nuclear resonance effect. Under ordinary conditions in a magnetic field, however, there is a slight excess of nuclei in the lower energy state. The principle involved in NMR then is the absorption of energy to convert nuclei from the lower to higher energy state. The difference between the energy states, ΔE, is proportional to the strength H of the applied magnetic field. It can be shown that $\Delta E = \gamma hH/2\pi$, where h is Planck's constant and γ is a proportionality constant typical of each kind of nucleus. In all kinds of spectroscopy involving absorption of electromagnetic radiation, ΔE is related to frequency, ν, of the radio frequency (Rf) oscillator by $\Delta E = h\nu$. Using the preceding two equations it is shown that $\nu = \gamma H/2\pi$. If ν is held constant and H varied until close to the point where $\nu = \gamma H/2\pi$, energy is absorbed by the nuclei, current flow from the oscillator increases, and may be detected. Further increases of H make $\nu < \gamma H/2\pi$ and the current decreases to its original value.

It is possible to hold H constant and vary the radio frequency or hold Rf constant and sweep the field with the external magnet. Most NMR spectrometers employ the latter technique. Those instruments on the market most often employ radio frequencies of 40 to 1000 megacycles (MHz).

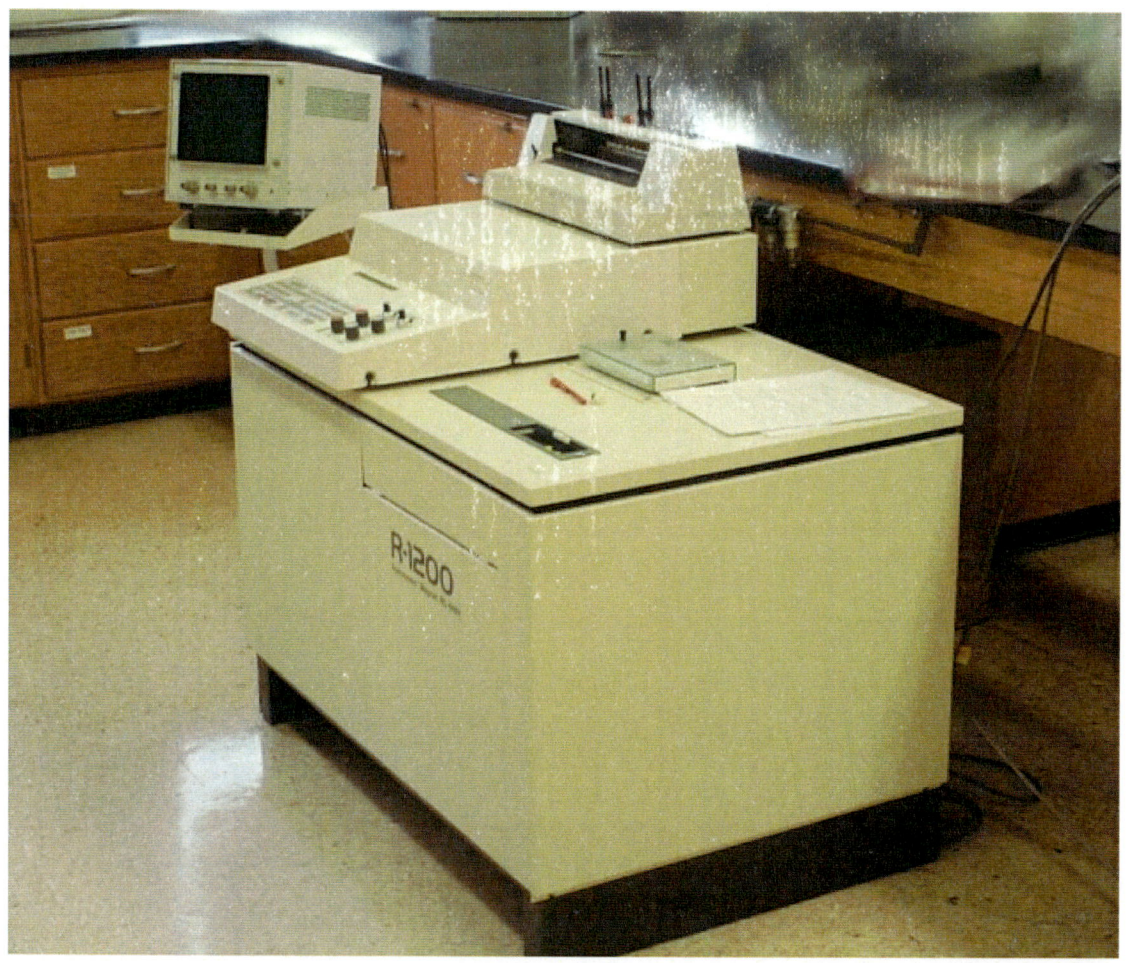

60 MHz Instrument

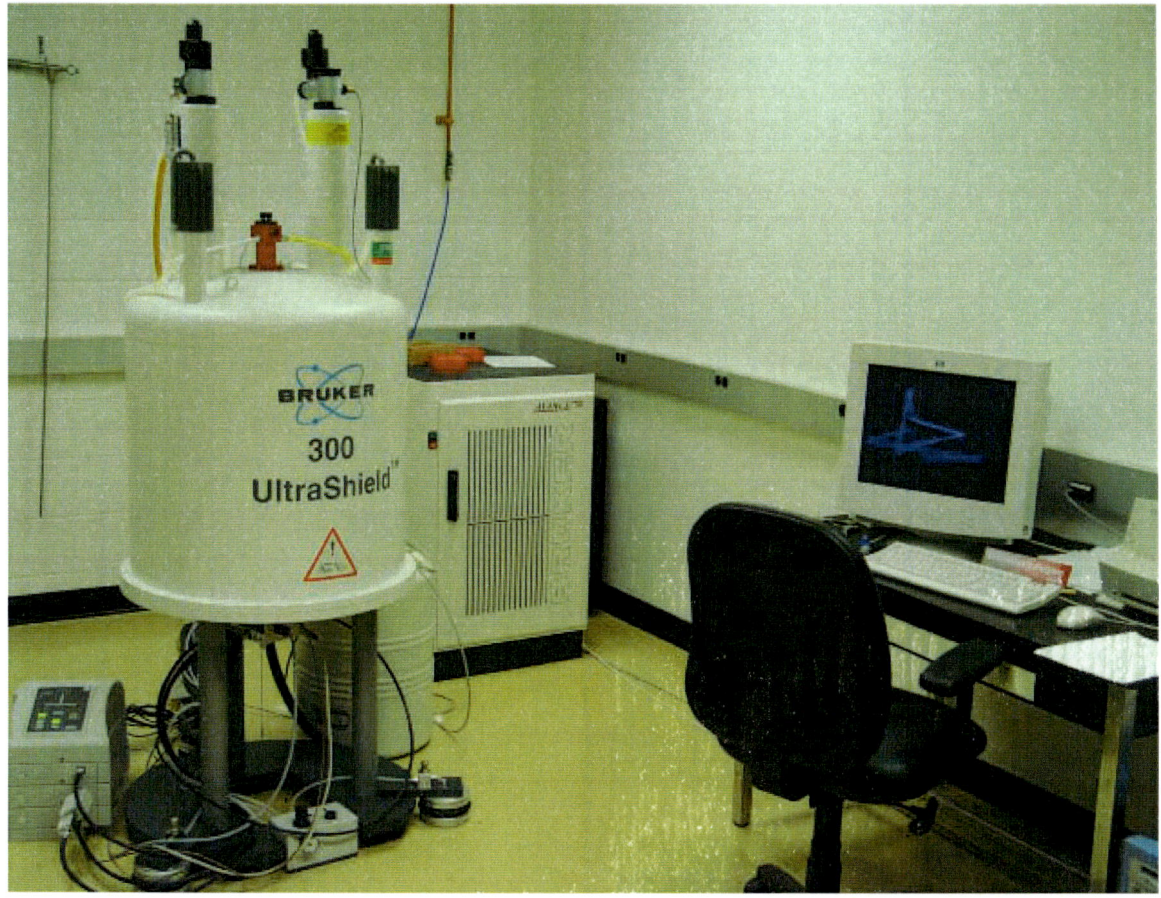

300 MHz NMR Spectrometer

Nuclear Magnetic Resonance spectroscopy is probably the most important technique used today for the structural study of organic and inorganic compounds. Substances are placed into a large magnetic field (commonly a superconducting magnet using a wire coil cooled to 4 K in a reservoir of liquid helium). The spinning nucleus of an atom either aligns with or against the externally applied magnetic field. Radio frequency radiation is used to "flip" the nuclear spin. The frequency absorbed by a nucleus is determined by the local electron density near each nucleus and by the geometry of the molecule. NMR spectra are used to determine protein structure and for countless other applications.

From the above discussion it might appear that all protons should absorb at the same place in the NMR spectrum. Fortunately, this is not true and NMR gives a method for differentiating among the various kinds of hydrogen in molecules. In a magnetic filed the nuclei (protons) of different kinds of hydrogen have slightly different magnetic environments. Surrounding electrons shield the nucleus so that the magnetic field actually felt by the nucleus is not quite the same as the applied field. Differences in the field strengths at which signals are obtained for nuclei of the same kind, but in different environments, are referred to as **chemical shifts.** In proton magnetic resonance and spectroscopy, tetramethylsilane (TMS) is usually the standard because there are very few protons in organic compounds that are more shielded than those in TMS. A small amount of TMS is dissolved in the sample to be studied. TMS shows up as a peak at the far right of the spectrum.

As a general rule it might be stated that protons in the environment of an electron-withdrawing group are less shielded and will absorb further downfield (to the left of TMS) than will those not in such an environment. Hence, the methylene hydrogens of bromoethane are less shielded than are methyl hydrogens.

Electronic shielding is proportional to the strength of the applied field and therefore, the chemical shift depends on field strength. In order to be able to express chemical shift independently of the field strength, the chemical shift parameter, δ or ppm, is used in reporting or analyzing a spectrum.

Aspects of The NMR Spectrum

The Signal: The signal intensity of each proton is the same. The area under each signal is proportional to the number of protons in the sample.

The Reference: NMR signals are reported relative to a reference signal. For protons, Tetramethylsilane (TMS, $Si(CH_3)_4$) is the reference compound. The protons in TMS appear at higher field than most others in conventional organic molecules. TMS absorbs at 0.00 ppm.

The Mechanics: In taking an NMR spectrum, the sample is dissolved in a solvent such as deuterium chloroform ($CDCl_3$), or deuterium acetone (CD_3COCD_3), or carbon tetrachloride (CCl_4) and a small amount of (1 %) of the reference, TMS is added. An acceptable spectrum may be routinely obtained on approximately 1 mg of material.

The NMR Scale: The position of the NMR signal relative to TMS is referred to as the chemical shift (δ) of that proton. δ is reported in parts per million (ppm).

NMR PROCEDURE

SAMPLE PREPARATION

1. Place one drop of liquid or about 1-10 mg of solid into a clean, dry vial.

2. Add about 0.5 ml of NMR solvent (usually $CDCl_3$ with TMS) to the tube. Make sure the sample dissolves. If $CDCl_3$ doesn't work, try D_6-acetone.

3. If you see particles, filter the solution by gravity.

4. Transfer solution into a clean, dry NMR tube.

5. Place a "turbine" on the tube and adjust to 10 cm from the bottom. There is a guage on the right side of the NMR instrument to help position the turbine.

DATE COLLECTION

6. Place the sample tube into the sample compartment, and make sure that it is spinning. Press the RESET button.

7. By pressing different WIDTH buttons, adjust the spectrum on the scope so you can see all of the peaks. Start at high settings (10 and 20 ppm indicators lit), go to a lower setting (only the 10 ppm indicator lit) if you can still see all of the peaks.

8. Turn the FIELD SHIFT knob to adjust the TMS peak (last on the right) to 1 ppm on the scope.

9. Set the balance knob between noon and three o'clock.

10. Set the magnetic field strength using the H_1 LEVEL controls. Make adjustments until the tallest peak is 3-4 blocks high.

11. Press the AUTO RES button and wait for the instrument to beep. This control fine tunes the magnetic field.

12. Press the ACQ SET button to set the sweep rate and filter settings to default values.

13. Hold down the ACQ button and press the ACQ START/STOP button to begin data acquisition.

DATA MANIPULATION AND PLOTTING

14. When data acquisition is finished, press the CORR button to smooth the spectrum a bit.

15. Adjust the balance knob so that the baseline on either side of each peak is level. Do your best; it probably won't be perfect.

    PLOTTER SETUP BEGINS HERE. IF THE SPECTROMETER WAS IN USE BEFORE YOU STARTED, SKIP TO STEP 17.

16. Make sure the plotter is on. (Check for lights.) If not, the on/off switch is on the left rear.

17. On the left side of the plotter set the paper switch to LOAD. Insert a piece of plotter paper and align it with the line that goes all the way across the plotter. Flip the paper switch to HOLD, and then hold down the SHIFT key and press the PAPER SET key on the plotter. This should load the paper.

THIS COMPLETES PLOTTER SETUP. RETURN TO THE INSTRUMENT KEYBOARD.

19. To set the reference, press the button in the middle of the arrows to activate the cursor on the scope. Hold down the right arrow until the cursor is at the top of the TMS peak. Press the left arrow if you overshoot the peak. Press REF then ENTER.

20. Press OFFSET twice to return the cursor to the original position. Then plot the scan by pressing the PLOTTER START/STOP button. Press SCALE to put the axes on the plot.

INTEGRATION

21. To integrate the scan press INT. Adjust the BALANCE knob until the integration line looks like level stair steps going up and to the right. Again, do your best; it won't be perfect.

22. To plot the integration press PLOTTER START/STOP.

CLEAN UP

23. Press RESET and remove your NMR tube from the instrument. Place the reference sample into the instrument.

24. Wash your NMR tube with acetone, then water, then acetone again and place them in the oven in the general chem lab. Remember that you are responsible for your NMR tube. You must clean it, recover it from the oven, and put it in your drawer for later use.

¹HNMR TABLE

Correlation of ¹H Chemical Shift with Environment

Type of proton	Formula	Chemical shift (δ)
Reference peak	$(CH_3)_4Si$	0
Saturated primary	$-CH_3$	0.7–1.3
Saturated secondary	$-CH_2-$	1.2–1.4
Saturated tertiary	$\diagdown\!\!-\!C\!-\!H\diagup$	1.4–1.7
Allylic primary	$\diagdown C\!=\!C\!-\!CH_3$	1.6–1.9
Methyl ketones	$\overset{\overset{\displaystyle O}{\|\|}}{-\,C\!-\!CH_3}$	2.1–2.4
Aromatic methyl	$Ar\!-\!CH_3$	2.5–2.7
Alkyl chloride	$Cl\!-\!C\!-\!H$	3.0–4.0
Alkyl bromide	$Br\!-\!C\!-\!H$	2.5–4.0
Alkyl iodide	$I\!-\!C\!-\!H$	2.0–4.0
Alcohol, ether	$-O\!-\!C\!-\!H$	3.3–4.0
Alkynyl	$-C\!\equiv\!C\!-\!H$	2.5–2.7
Vinylic	$\diagdown C\!=\!C\!-\!H\diagup$	5.0–6.5
Aromatic	$Ar\!-\!H$	6.5–8.0
Aldehyde	$\overset{\overset{\displaystyle O}{\|\|}}{-C\!-\!H}$	9.7–10.0
Carboxylic acid	$\overset{\overset{\displaystyle O}{\|\|}}{-C\!-\!O\!-\!H}$	11.0–12.0
Alcohol	$\diagdown\!\!-\!C\!-\!O\!-\!H\diagup$	Extremely variable (2.5–5.0)

APPROXIMATE CHEMICAL SHIFT RANGES (PPM) FOR SELECTED TYPES OF PROTONS[a]

Structure		Shift	Structure		Shift
R–CH₃		0.7–1.3	R–N–C–H		2.2–2.9
R–CH₂–R		1.2–1.4	R–S–C–H		2.0–3.0
R₃CH		1.4–1.7	I–C–H		2.0–4.0
R–C=C–C–H		1.6–2.6	Br–C–H		2.7–4.1
R–C(O)–C–H, H–C(O)–C–H		2.1–2.4	Cl–C–H		3.1–4.1
RO–C(O)–C–H, HO–C(O)–C–H		2.1–2.5	R–S(O)(O)–O–C–H		Ca–3.0
N≡C–C–H		2.1–3.0	RO–C–H, HO–C–H		3.2–3.8
(phenyl)–C–H		2.3–2.7	R–C(O)–O–C–H		3.5–4.8
R–C≡C–H		1.7–2.7	O₂N–C–H		4.1–4.3
R–S–H	Var	1.0–4.0[b]	F–C–H		4.2–4.8
R–N–H	var	0.5–4.0[b]			
R–O–H	var	0.5–5.0[b]			
(phenyl)–O–H	var	4.0–7.0[b]	F–C=C–H		4.5–6.5
(phenyl)–N–H	var	3.0–5.0[b]	(phenyl)–H		6.5–8.0
R–C(O)–N–H	var	5.0–9.0[b]	R–C(O)–H		9.0–10.0
			R–C(O)–OH		11.0–12.0

[a]For those hydrogens show as –C–H, if that hydrogen is part of a methyl group (CH₃) the shift is generally at the low end of the range give, if the hydrogen is in a methylene group (–CH₂–) the shift is intermediate, and if the hydrogen is in a methine group (–CH–), the shift is typically at the high end of the range given.

[b]The chemical shift of these groups is variable, depending not only on the chemical environment in the molecule, but also on concentration, temperature, and solvent.

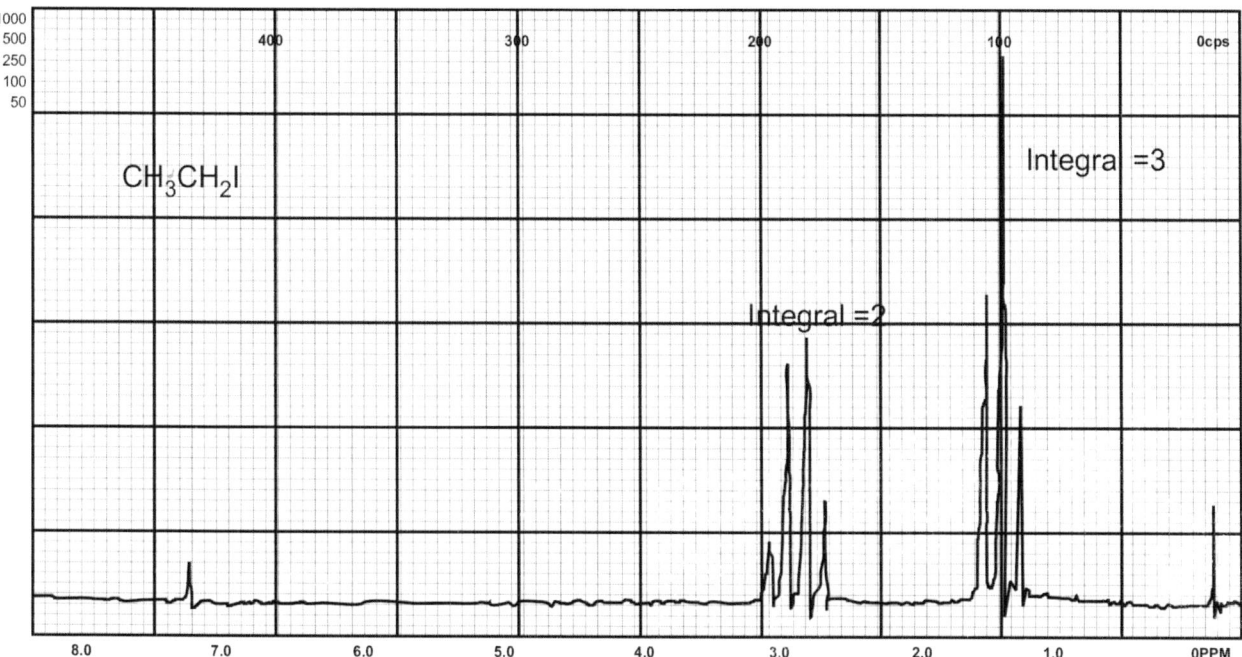

CH₃CH₂I

Integral =3

Integral =2

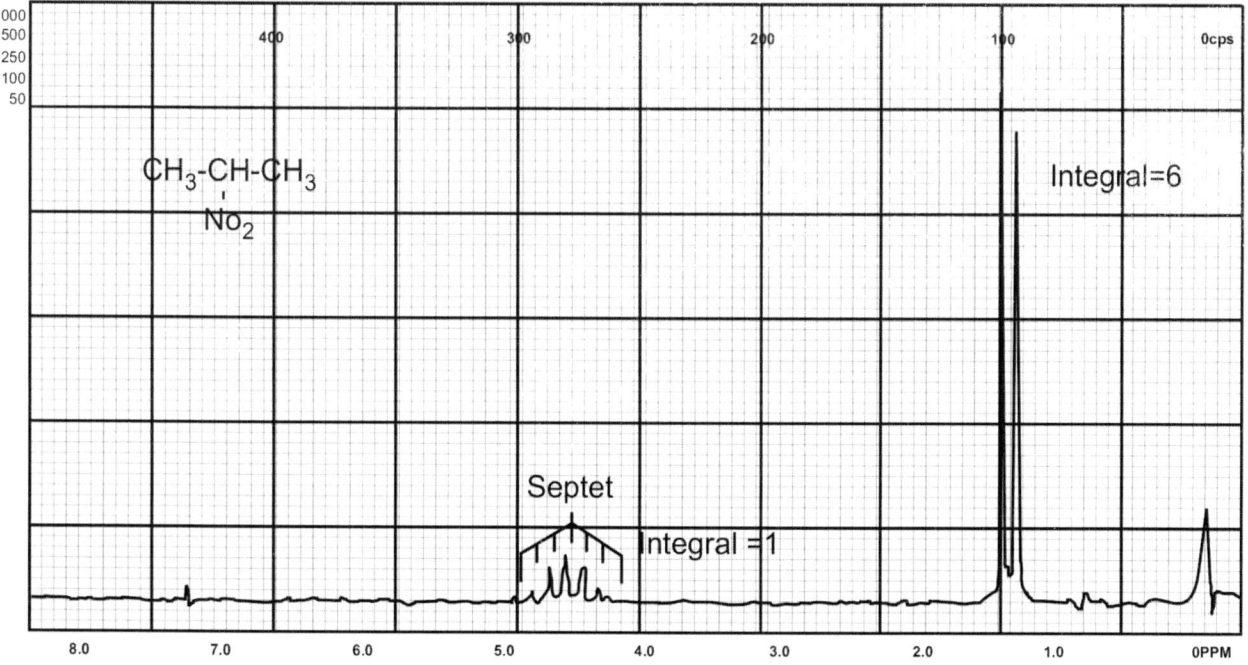

CH₃-CH-CH₃
|
NO₂

Integral=6

Septet

Integral =1

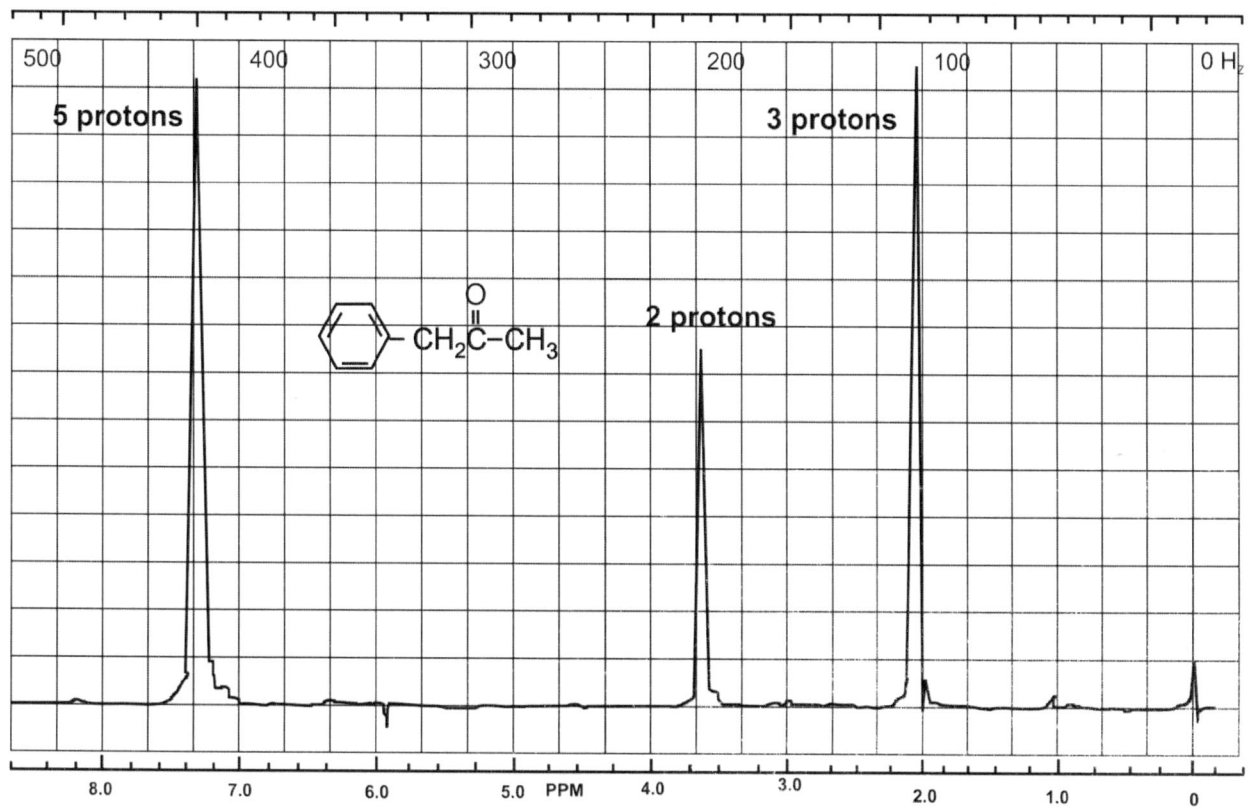

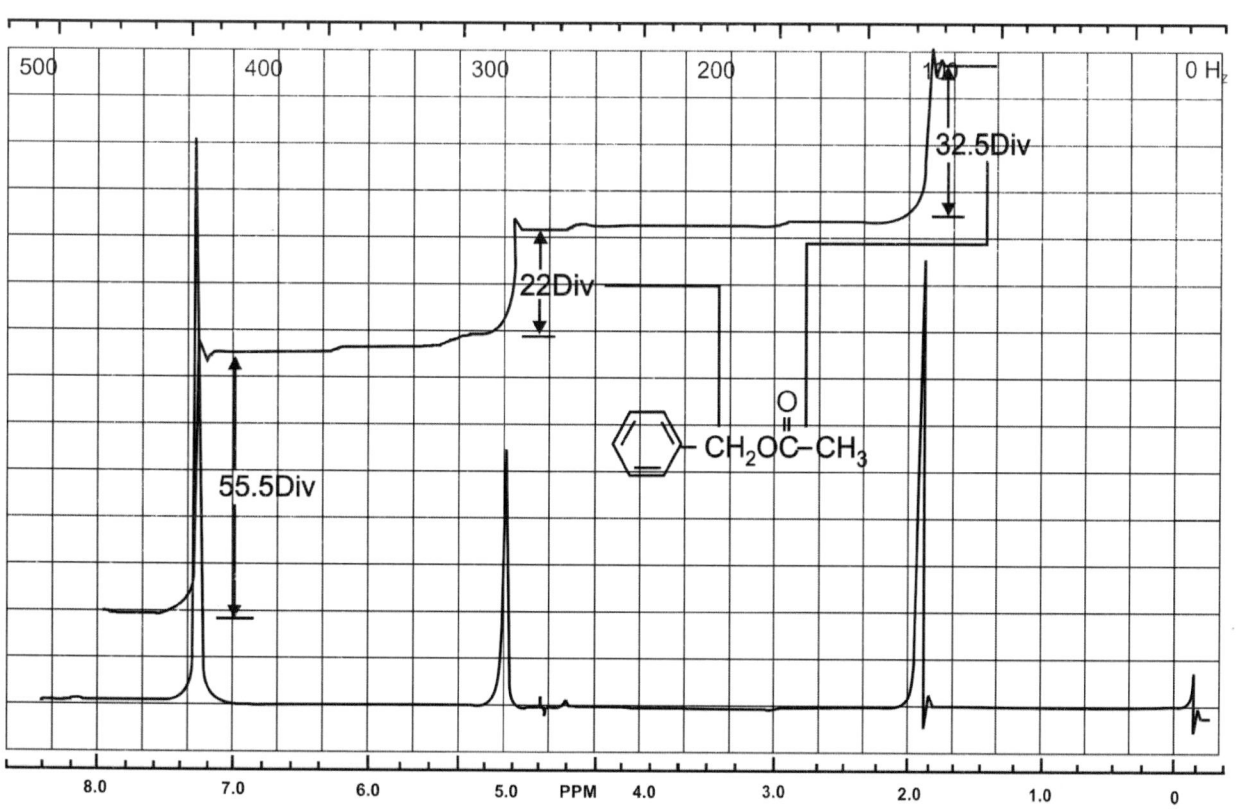

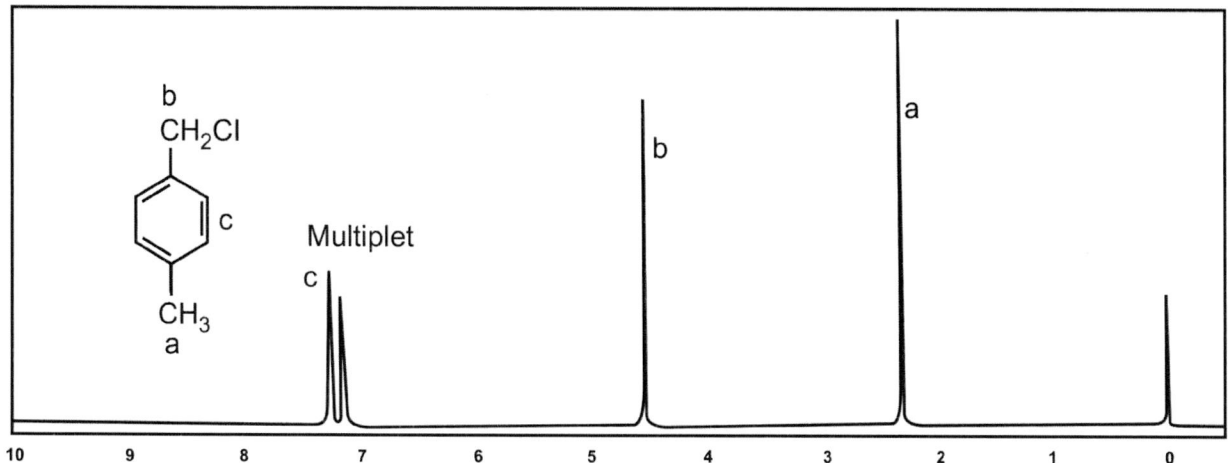

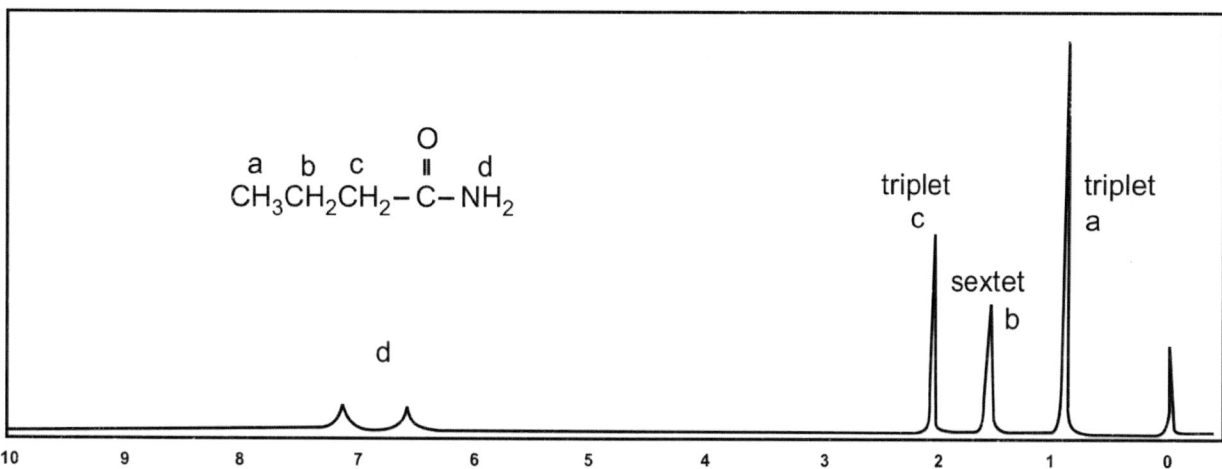

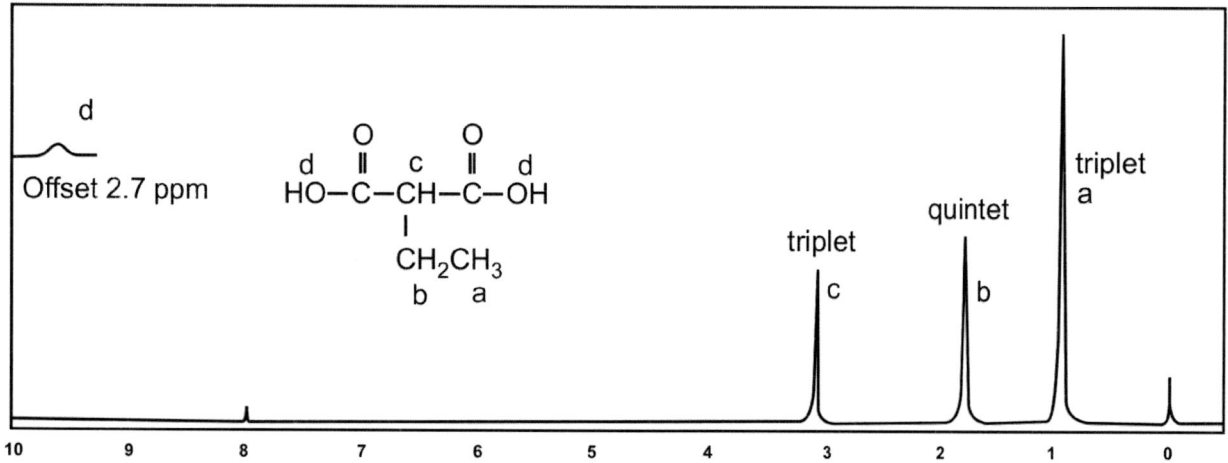

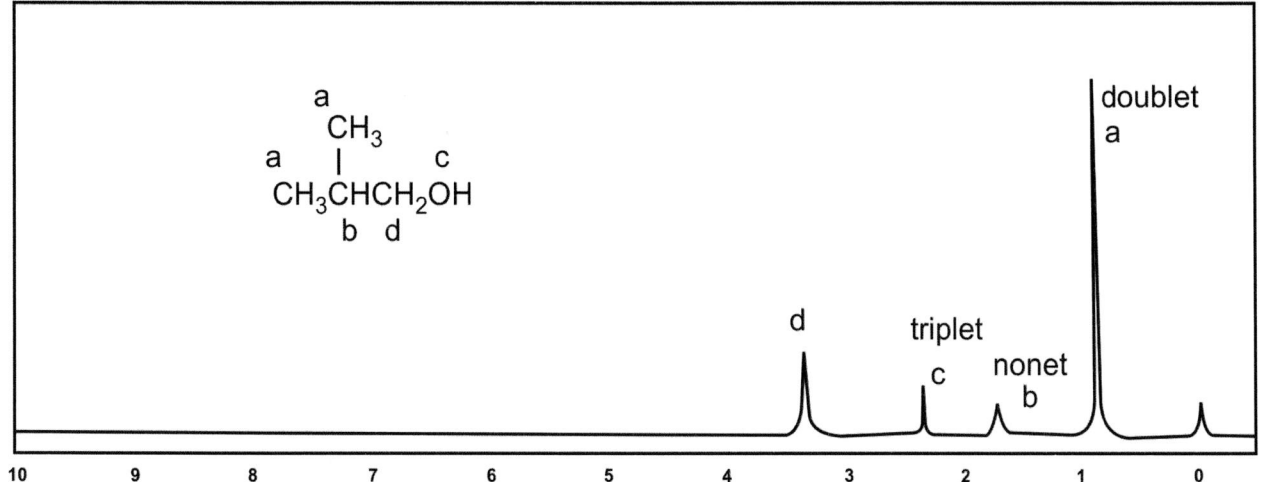

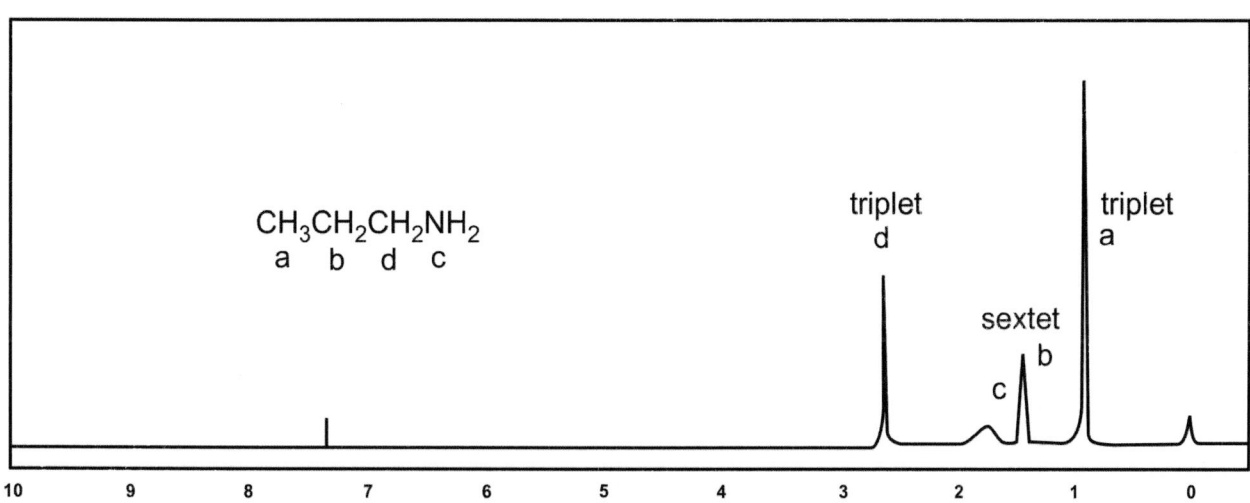

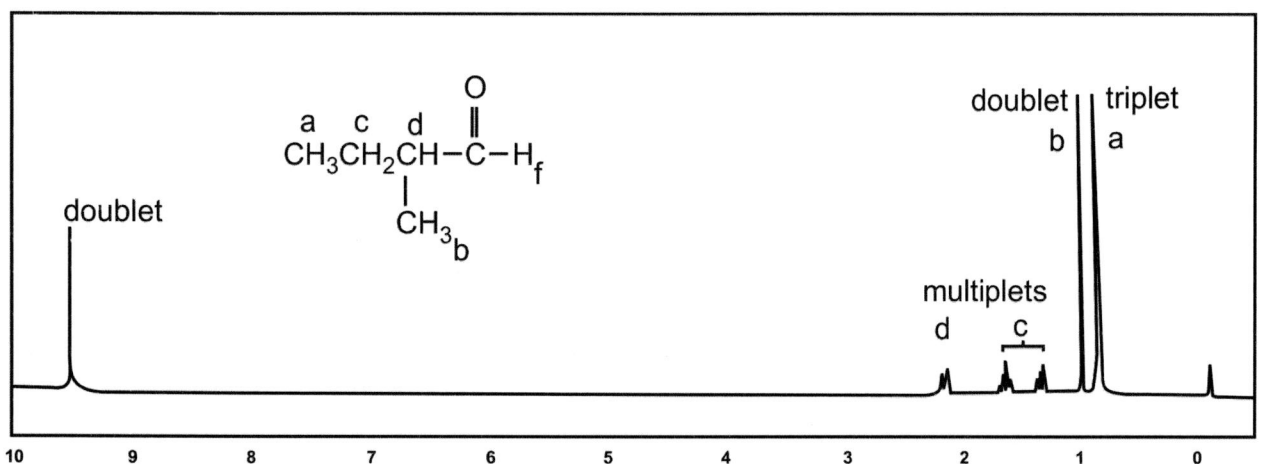

COMBINED IR AND NMR EXAMPLES

$C_9H_{12}O$

P 5360-8   2-n-Propylphenol
$CH_3CH_2CH_2C_6H_4OH$ M.W. 136.19 n_D^{20} 1.5224
b.P. 224.6-226.6°..............................

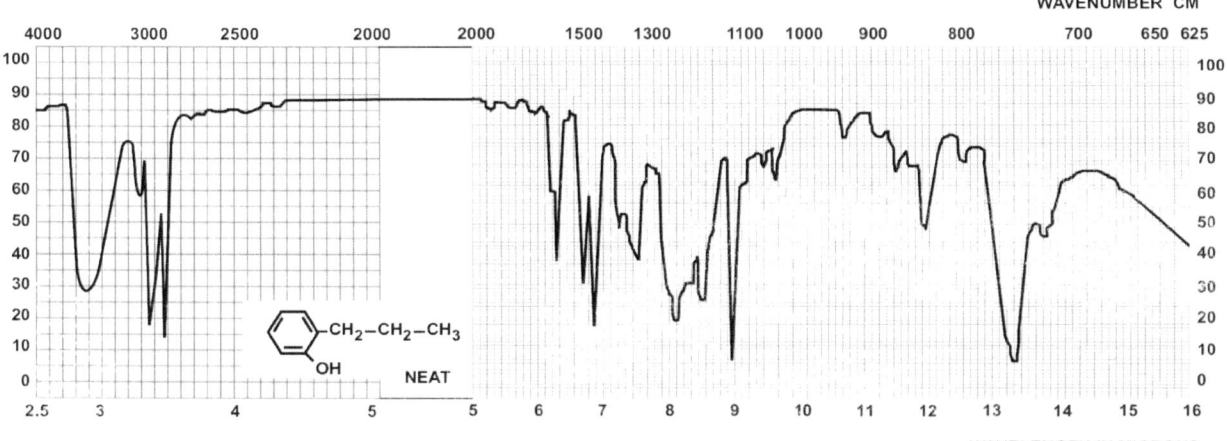

P5,360-8   2--n-Propylphenol, 98%
$CH_3CH_2CH_2C_6H_4OH$ M.W. 136.19 d.p. 224-226°
n_D^{20} 1.5279 Beil. 6,499 IR 486D

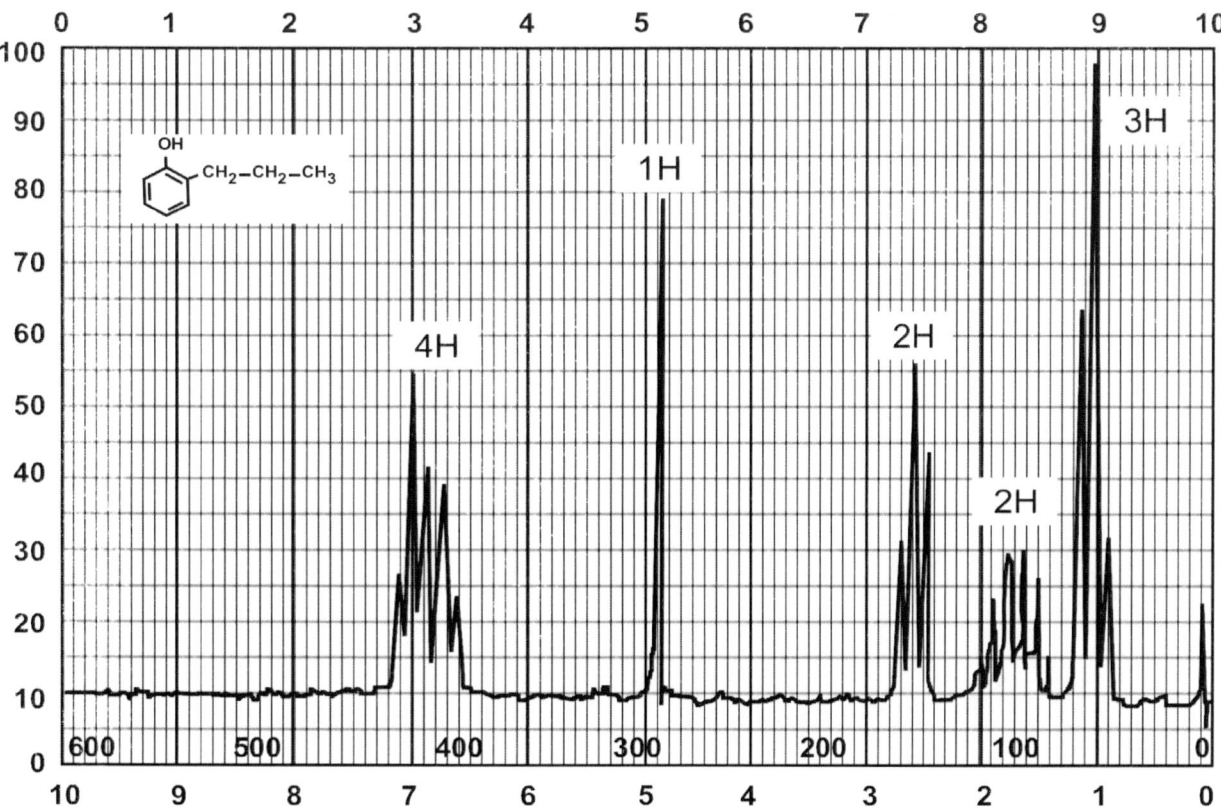

$C_{10}H_{14}O$

13.528-3   410C.- Butylphenol
$C_2H_5CH(CH_3)C_6H_4OH$ M.W. 150.22
m.p. 54-58⁰.............................

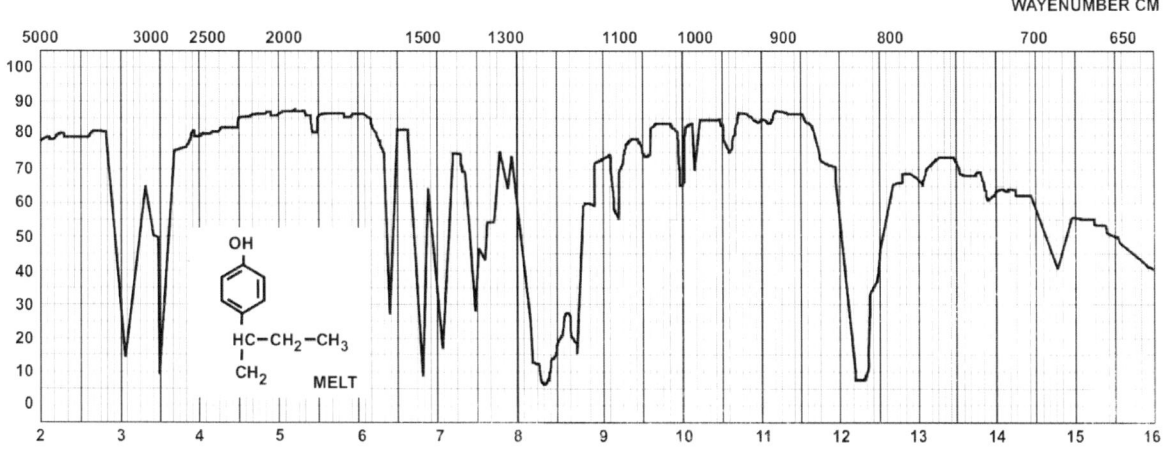

13,528-3
4-Sθc-Butylphenol
$C_2H_5CH(CH_3)C_6H_4OH$
M.W. 150.22 b.p. 135-136⁰/25mm. Beil. 6.522
IR 489D IRRITANT

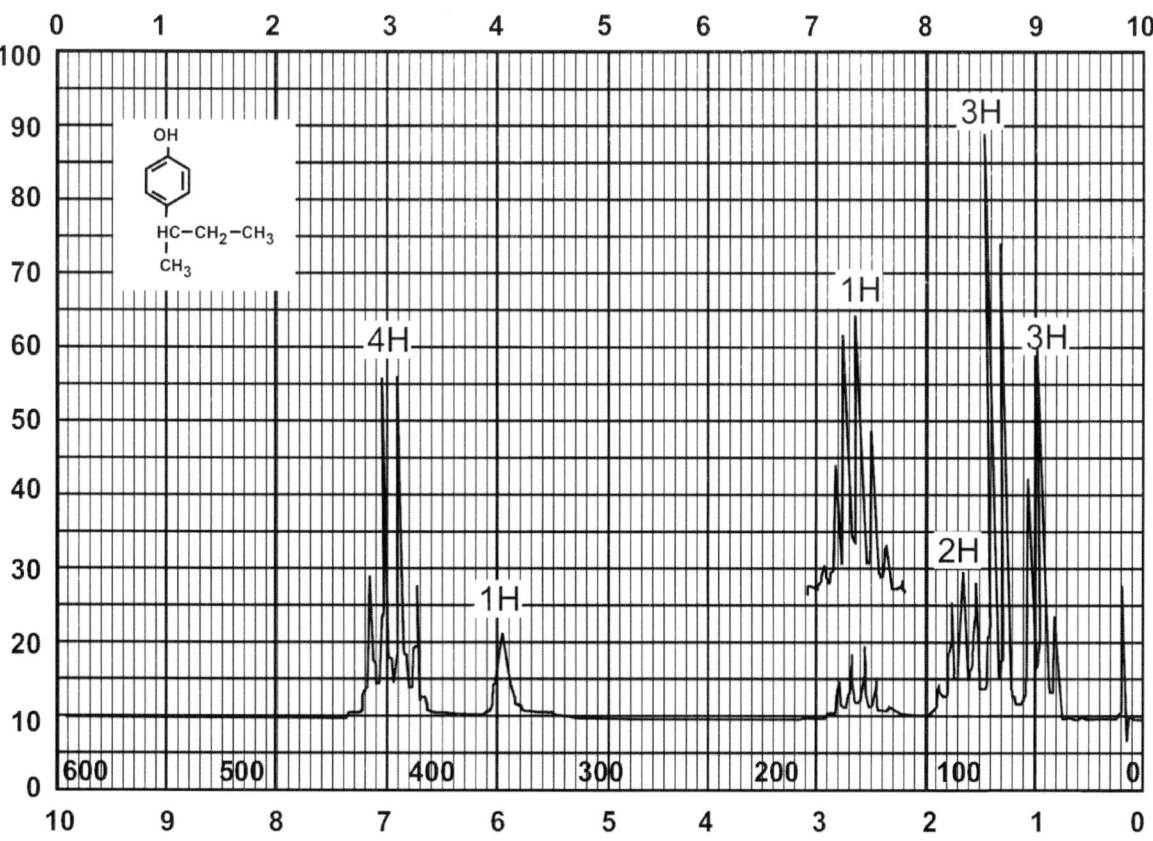

C₅H₈O₂

$$C_5H_8O_2$$

D13,860-6 3,3 - Dimethylacrylic acid (3-methyl-2-butenoic acid)
     $(CH_3)_2$ C:CHCO$_2$H M.W. 100.12
     m.p 68.5-69.5⁰.............................

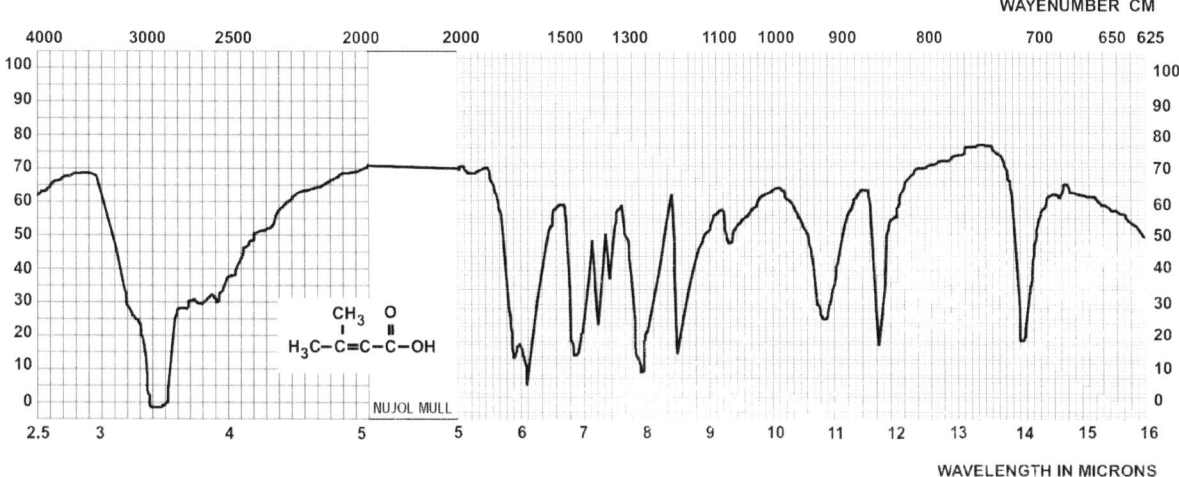

D13,860-6 3,3 -
Dimethylacrylic acid 97% (3-methyl-2-butenoic acid, senecioic acid)
$(CH_3)_2$ C=CHCO$_2$ M.W. 100.12 m.p. 68.5-69.5⁰ b.p. 194-195⁰
Beil. 2,435 IR 230F IRRITANT (send for data sheet)

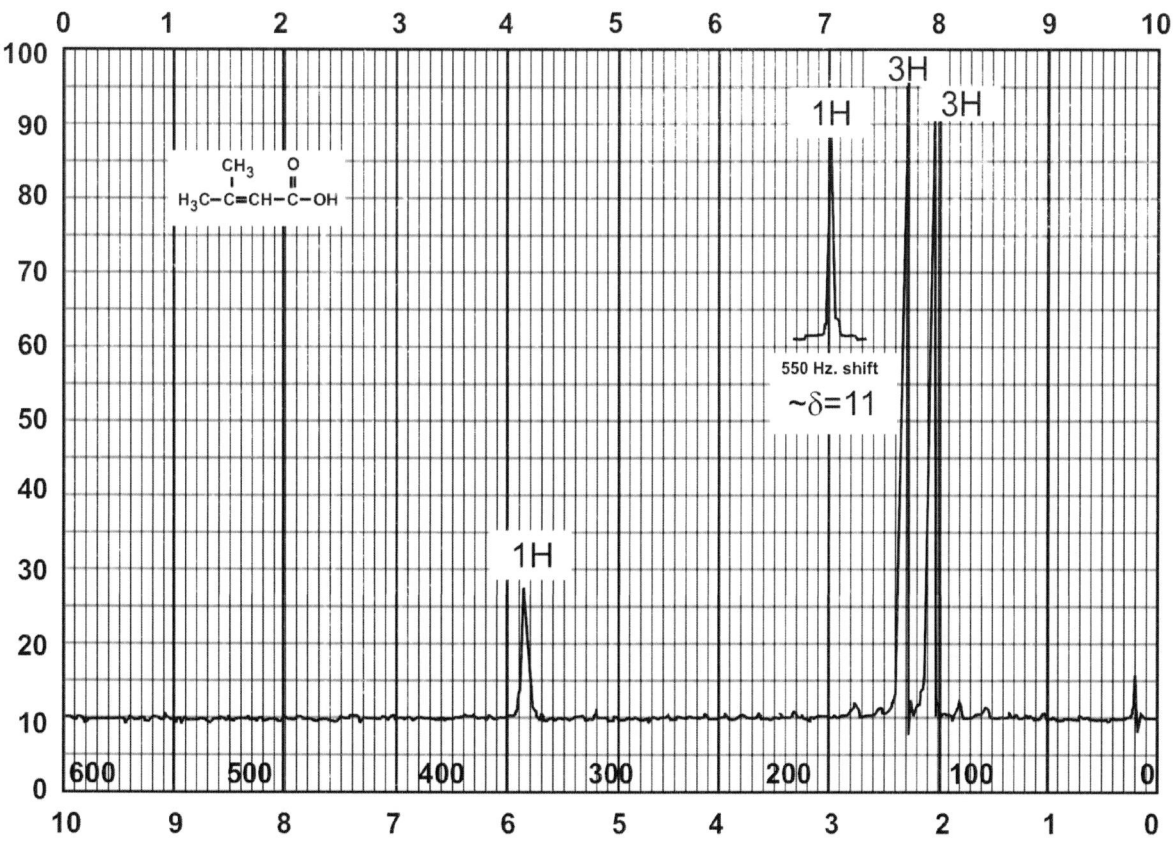

$C_6H_{12}O_2Cl_2$

D 5420-6   Dichloroacetaldehyde diethyl acetal (1,1-dichloro-2,2-diethoxyethane)

$Cl_2CHCH(OC_2H_5)_2$ M.W. 187.07............................

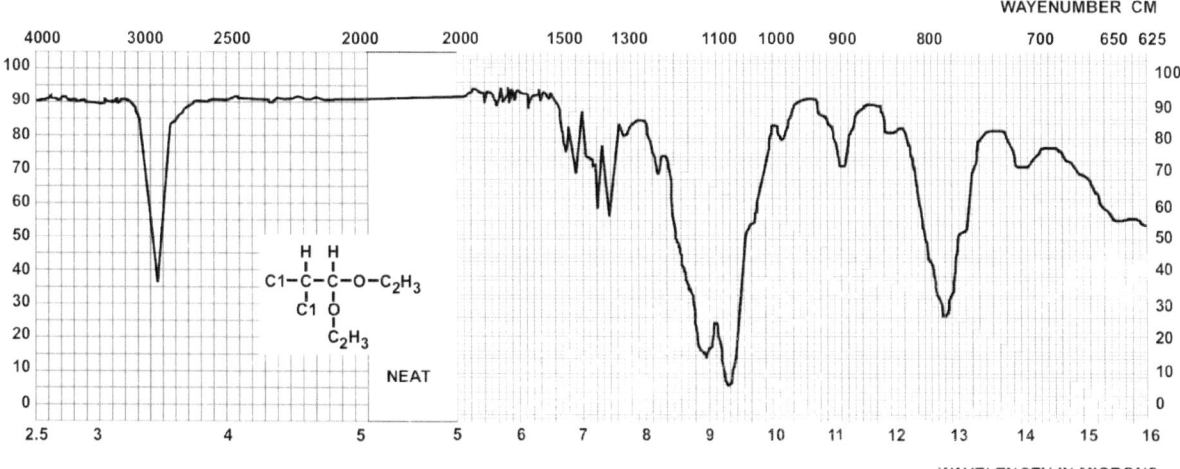

D 5,420-6

Dichloroacetaldehyde diethyl acetal 99%

(1,1-dichloro-2,2-diethoxyethane)

$Cl_2CHCH(OC_2H_5)_2$

M.w. 187.07 b.p. 183-184° n_D^{20} 1.4360 d 1.138 Beil. 1.614 IR 104B IRRITANT

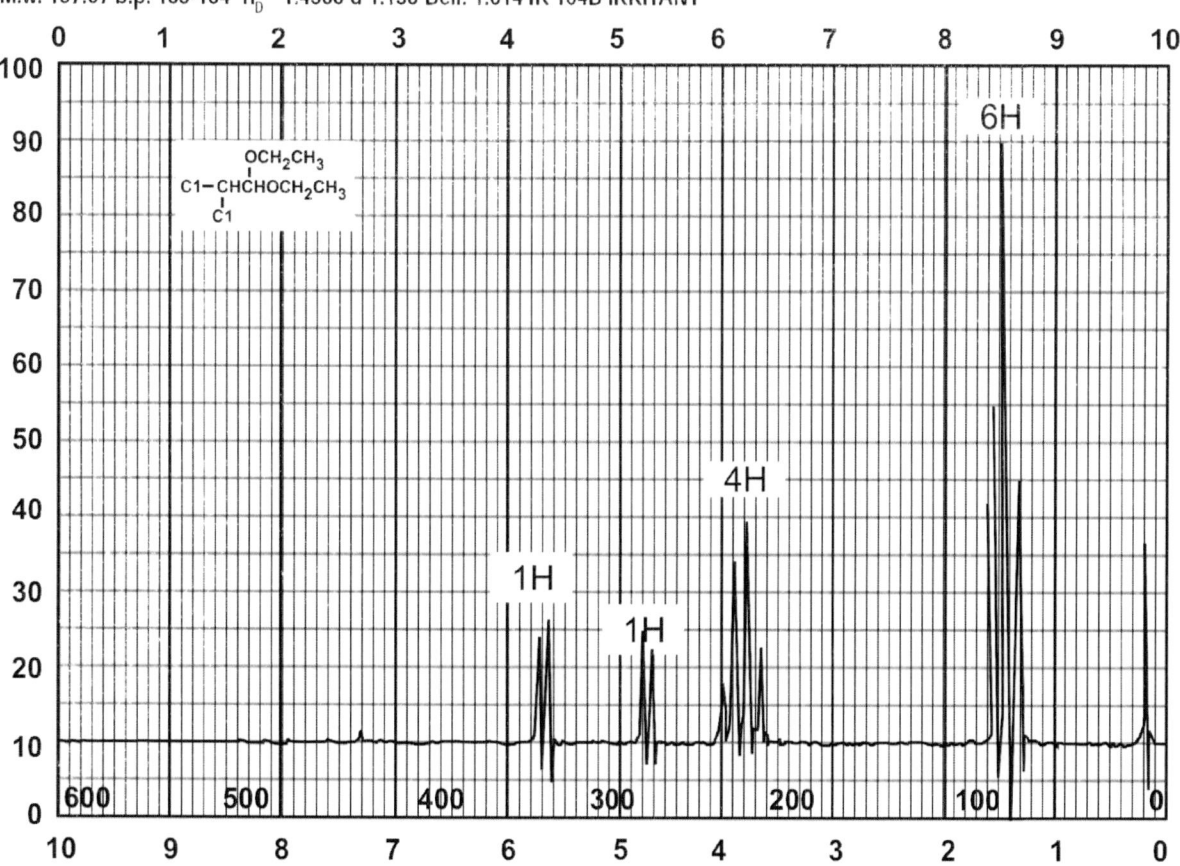

$C_4H_6Cl_2$

14,540-8 1,3-Dichloro-2-butene
    $CH_3C(Cl){:}CHCH_2Cl$ M.W. 125.01
    n_D^{20} 1.4720.............................

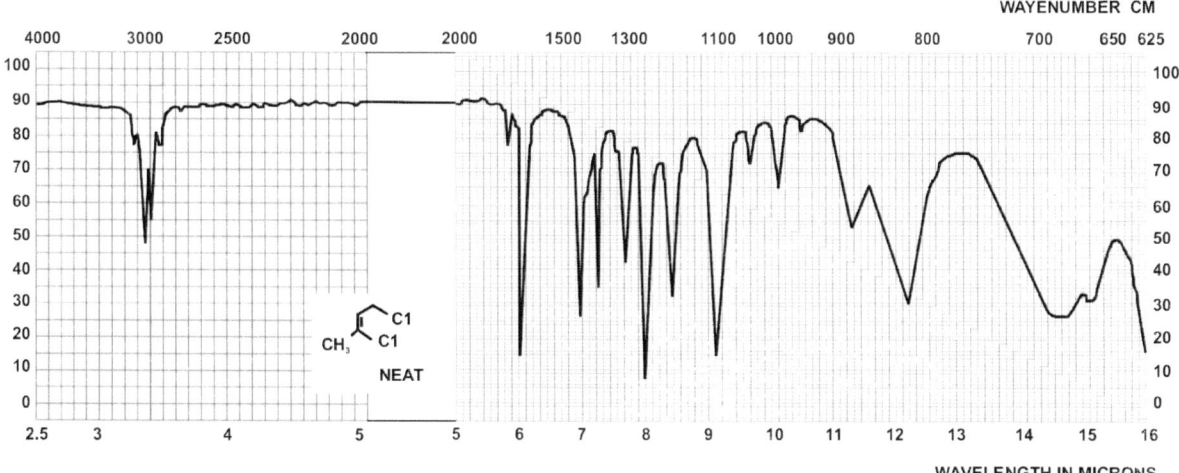

WAYENUMBER CM

WAVELENGTH IN MICRONS

14,540-8
1,3-Dichloro-2-butene, 98%
$CH_3C(Cl){=}CHCH_2Cl$
M.W. 125 n_D^{20} 1.4720 d 1.154 Fieser 1,2142,111
IR 52D LACHRYMATOR

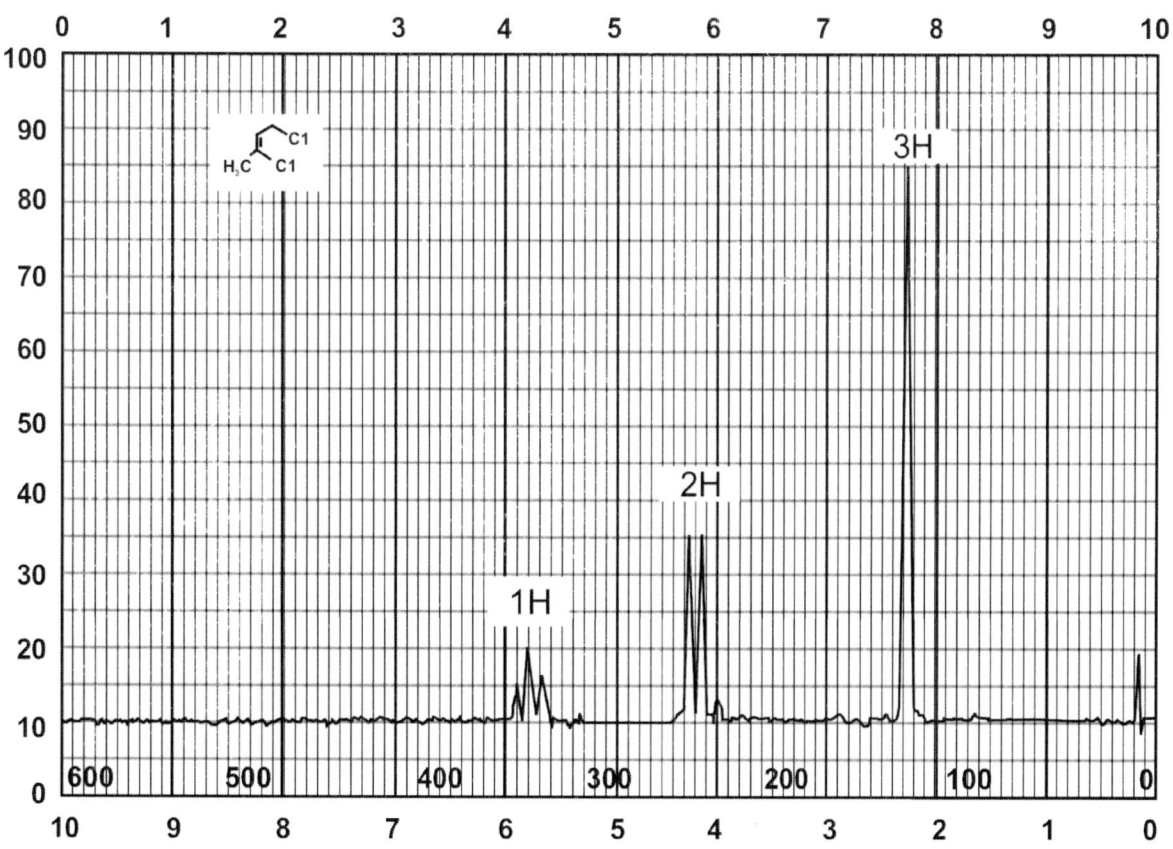

C₈H₁₀O

$$C_8H_{10}O$$

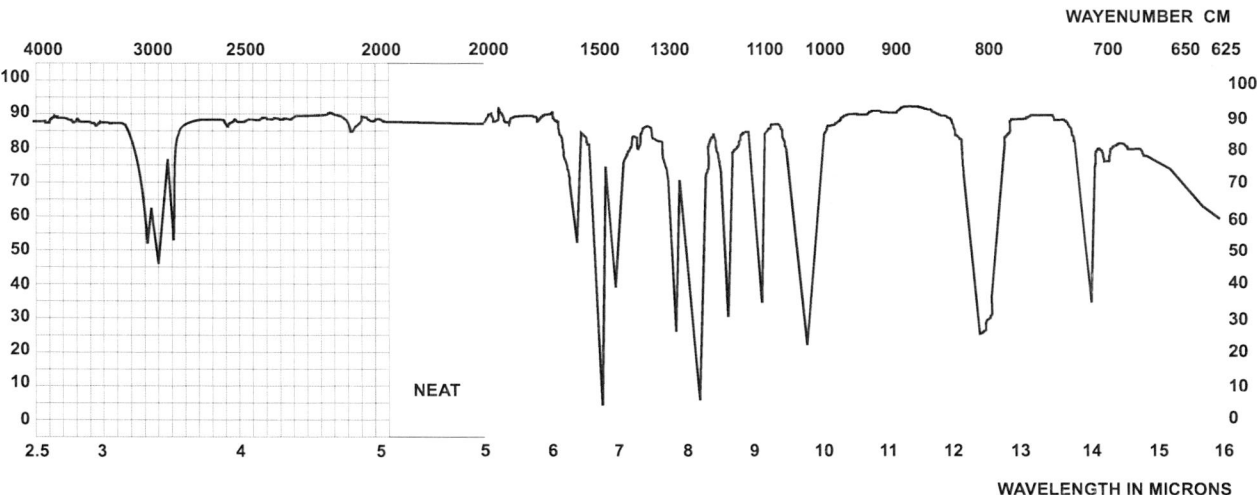

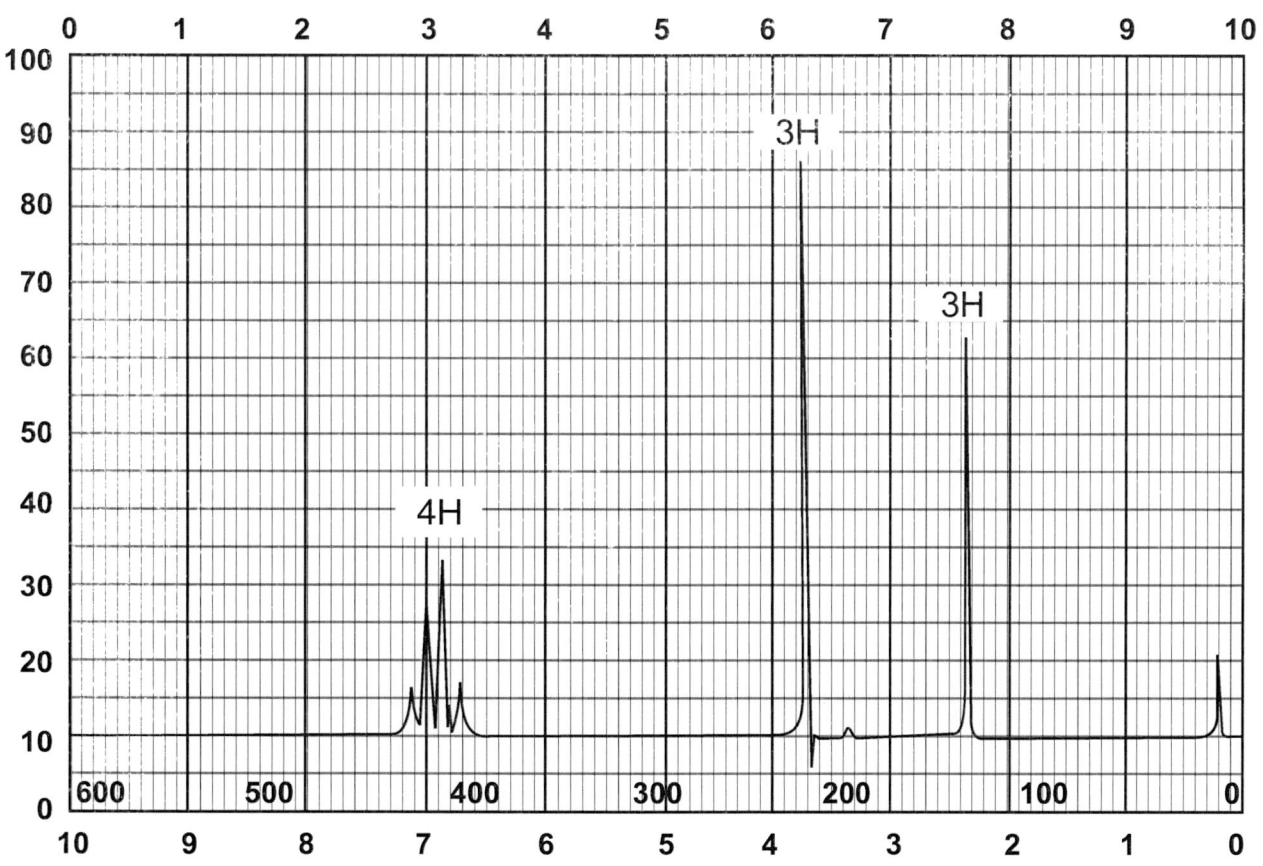

$C_6H_{14}O_2$

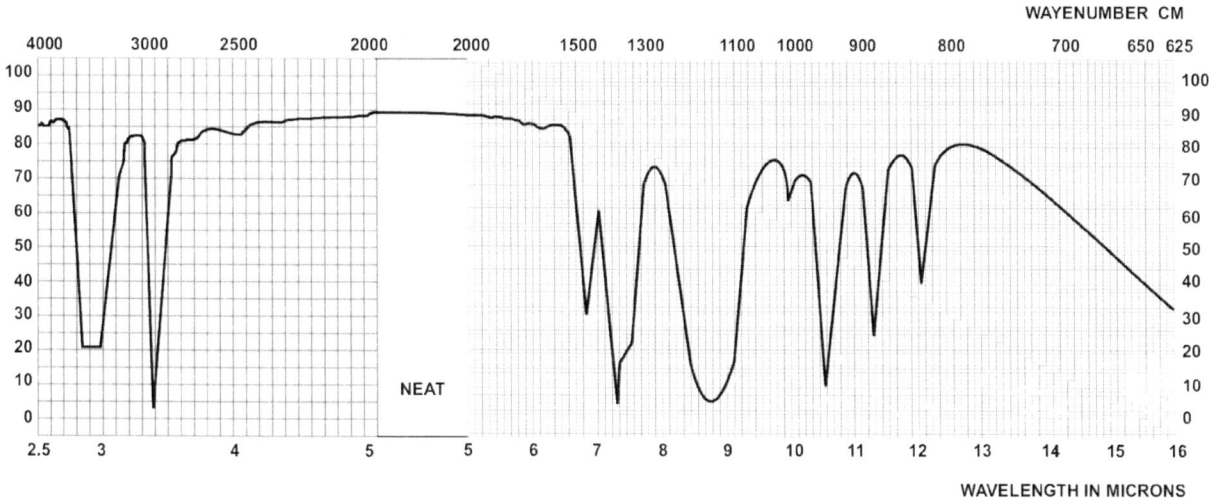

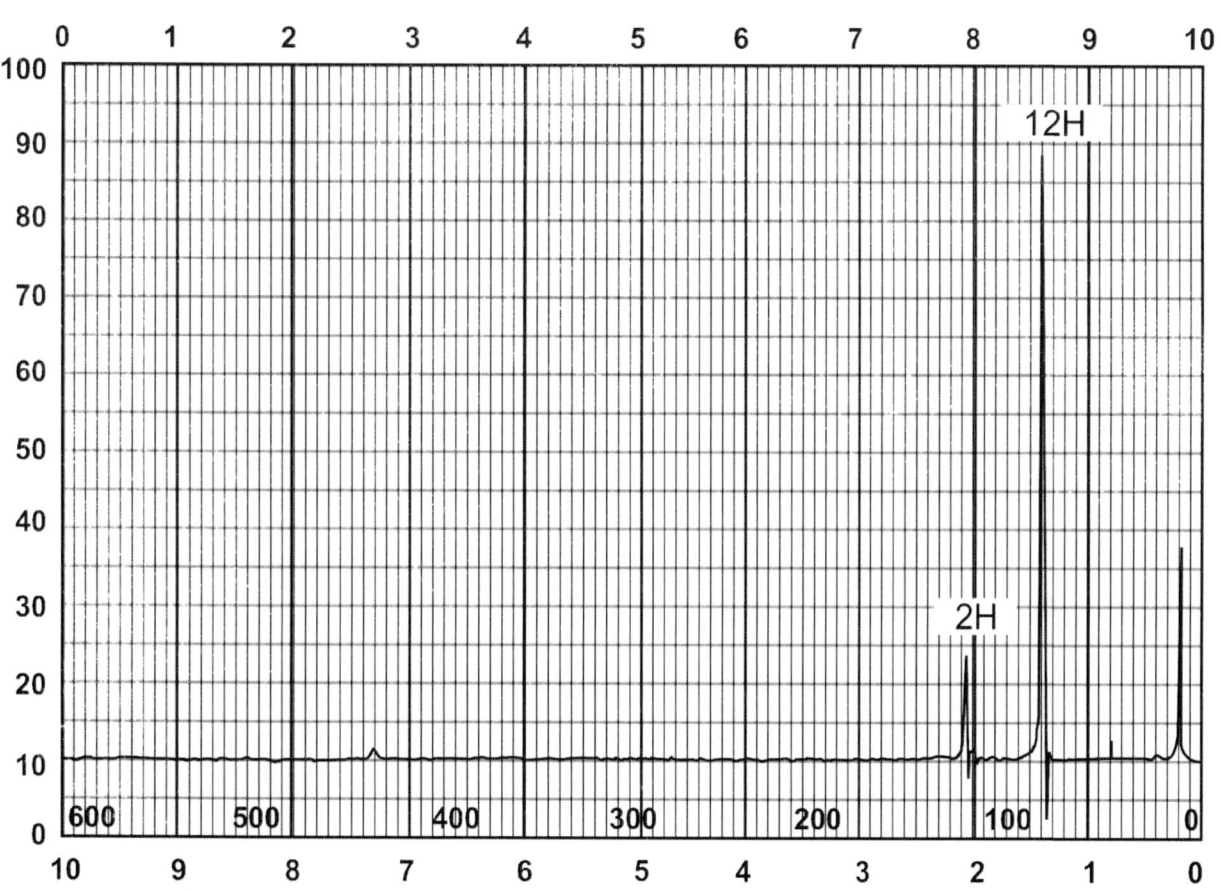

C$_3$H$_7$OCl

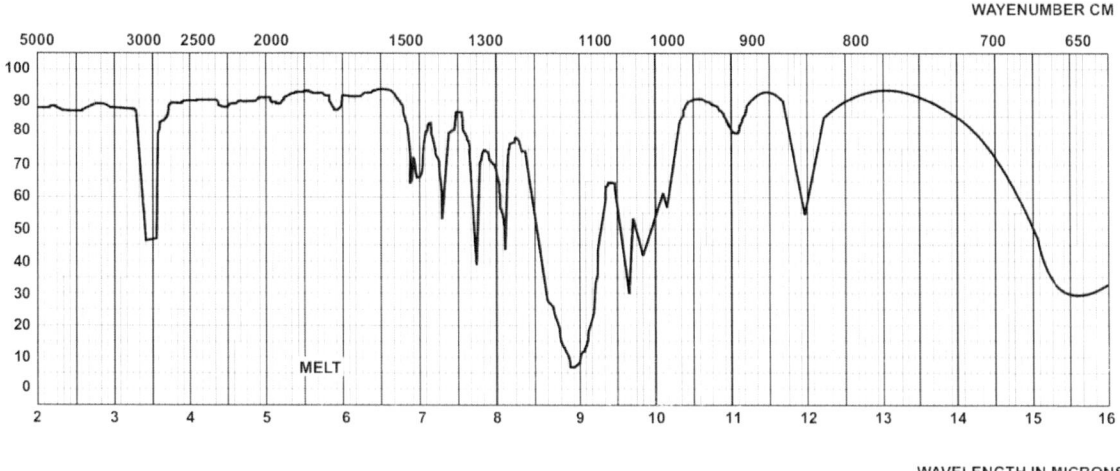

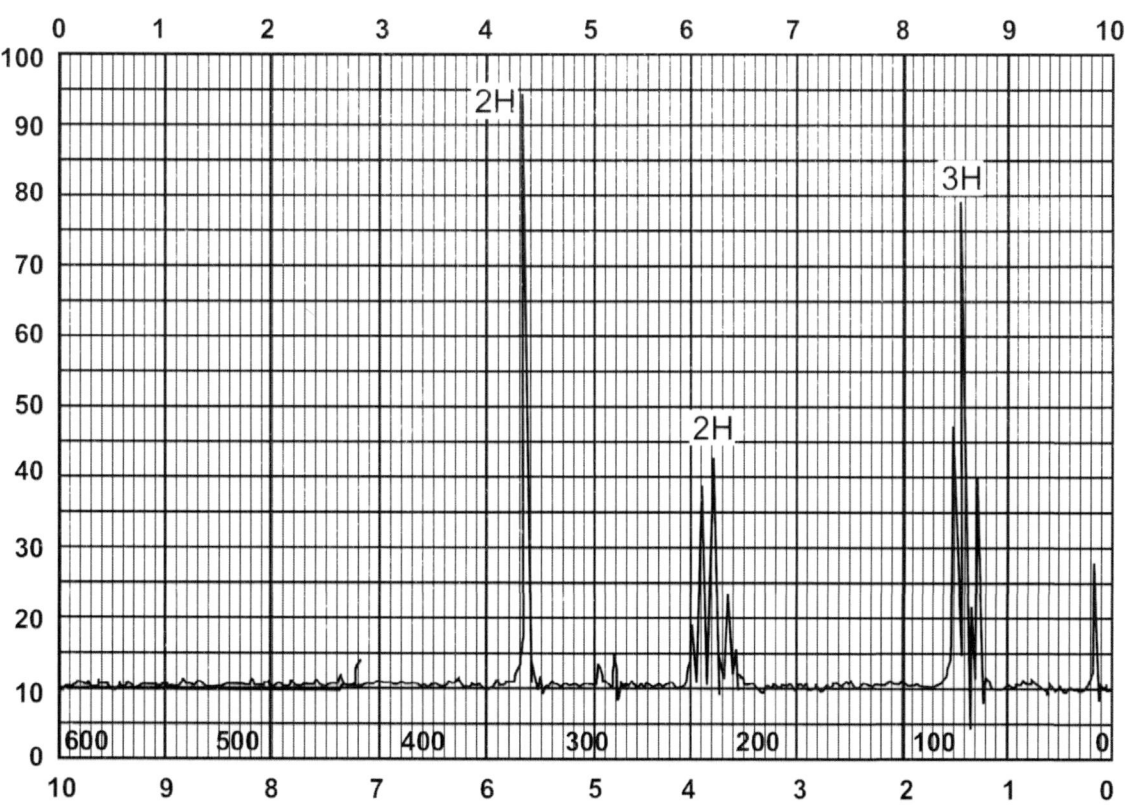

$C_5H_{10}O_2$

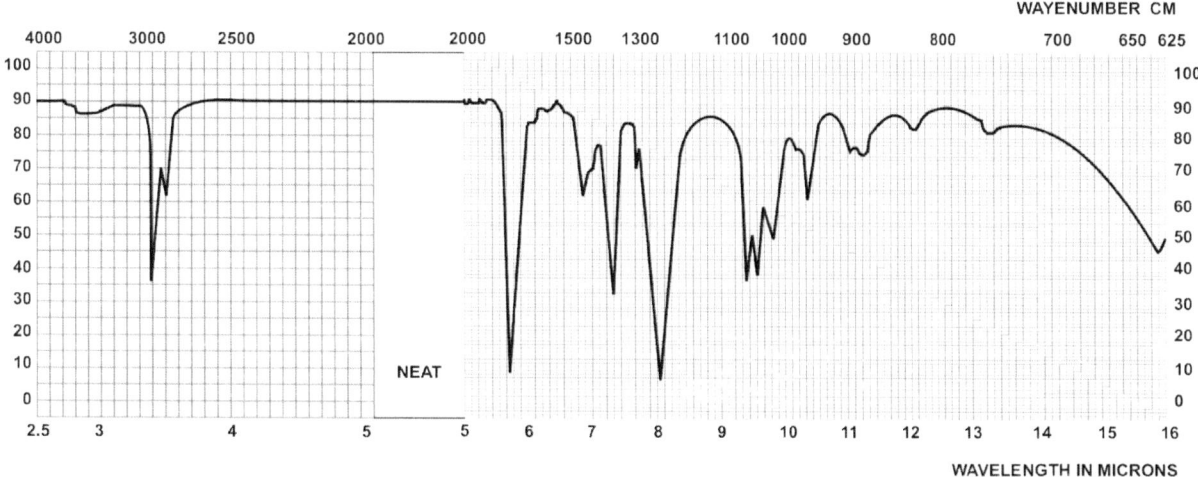

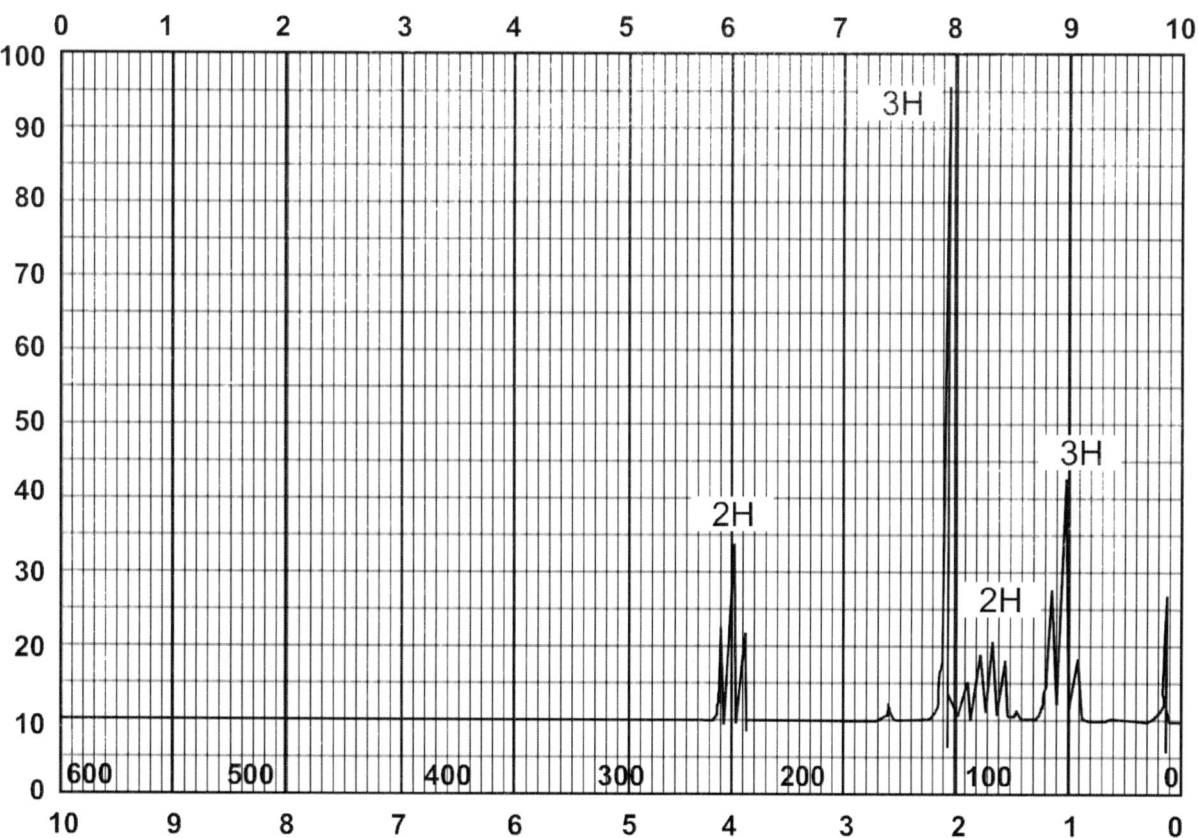

C$_5$H$_{14}$N$_2$

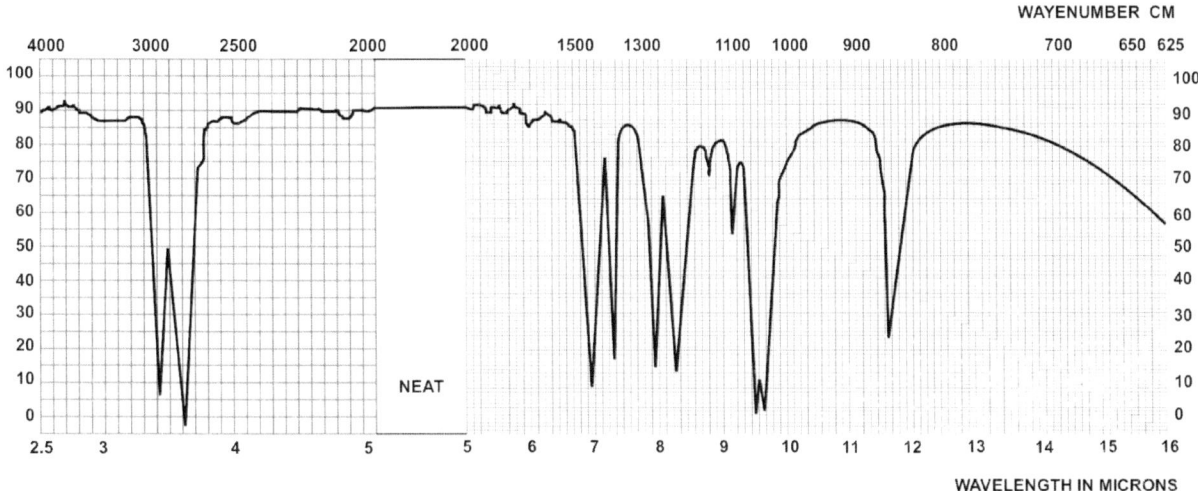

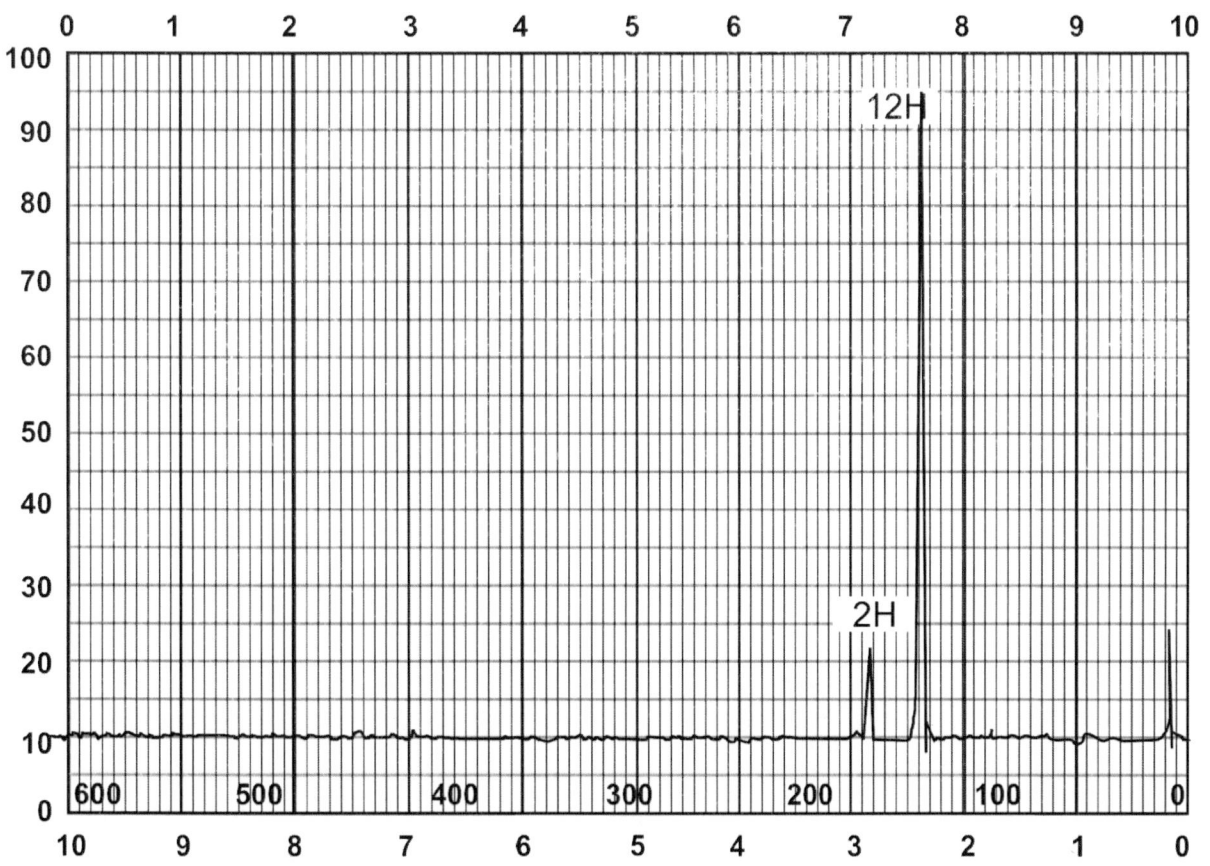

C_4H_8O

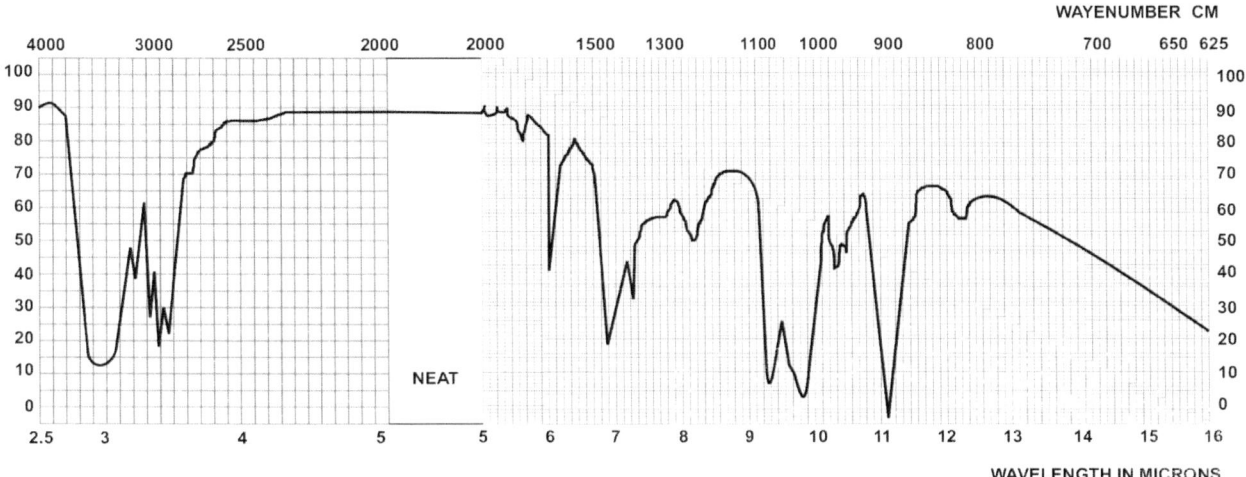

NEAT

WAYENUMBER CM

WAVELENGTH IN MICRONS

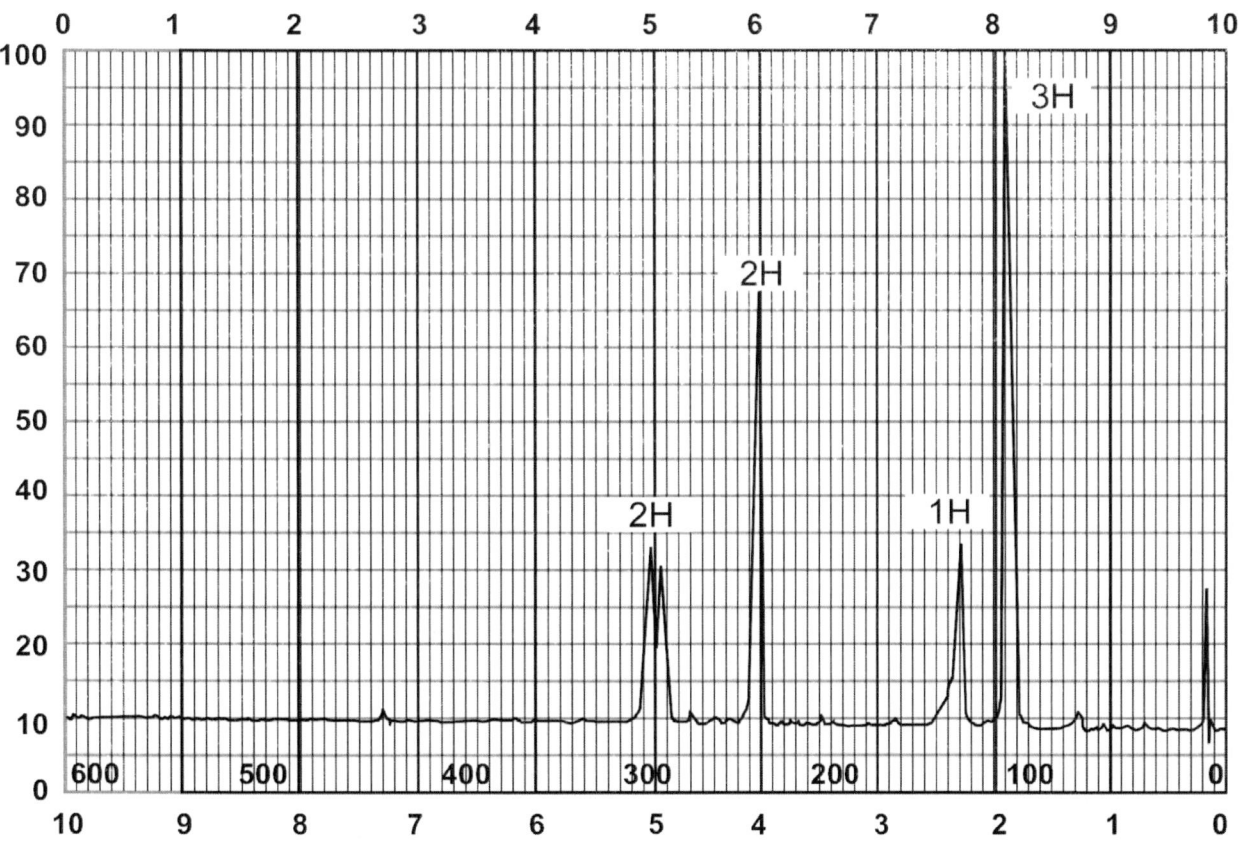

Carbon – 13 NMR Spectrometry

Nuclear magnetic resonance of carbon and hydrogen nuclei share some characteristics but there are several important differences. The most abundant isotope of carbon is carbon – 12. This isotope of carbon has no magnetic moment and is invisible in an NMR experiment. An isotope of carbon that does have a magnetic moment (spin quantum number = ½) is carbon – 13. The natural abundance of ^{13}C is only 1.1% and this has several important consequences:

1. The occurrence of a $^{13}C – ^{13}C$ bond is rare and coupling between ^{13}C nuclei, which would complicate spectra, is not observed.

2. ^{1}H NMR spectra are not complicated by $^{13}C – ^{1}H$ coupling.

3. ^{1}C NMR spectra are complicated by $^{13}C – ^{1}H$ coupling. This provides information about the number of hydrogen that reside on a given carbon.

A disadvantage of the low natural abundance of ^{13}C is sensitivity. Furthermore, the sensitivity of a nucleus in an NMR experiment is proportional to the gyromagnetic ratio of the nucleus (γ) and γ for ^{13}C is only one fourth that of a ^{1}H. Thus, the overall sensitivity of the ^{13}C relative to a ^{1}H is about 1/6000. Nonetheless, with the advent of Fourier Transform NMR (FT – NRM) it became possible to obtain ^{13}C spectra.

Important Information on ^{13}C – NMR

1. ^{1}H- Decoupled ^{13}C – NMR

- Gives information on the number and type of carbons.
- Each carbon appears as a single peak with a characteristic chemical shift.

2.   Off-Resonance Decoupled ^{13}C – NMR

- Deals with ^{13}C – ^{1}H Coupling
- gives the number of protons (hydrogen) on a given carbon
- Spin-spin splitting just like in ^{1}H NMR
- Can be very difficult to interpret.

3.   Dept ^{13}C – NMR

- Gives information of type of carbons and the number of hydrogen attached to each carbon.
- Process occurs in three stages.

   a.   Broadband decoupled – shows all carbons

   b.   Dept 90 – deals with CH (positive)

   c.   Dept 135 – deals with CH and CH3 (positive) and CH2 (negative)

^{13}C NMR: Chemical Shifts

The factors governing chemical shift in ^{13}C NMR spectra are largely the same as those governing chemical shift in ^{1}H NMR spectra. Thus, the more electron deficient the carbon atom, the further downfield the chemical shift (relative to TMS). Empirically derived parameters have been developed for estimating the chemical shift expected for a given ^{13}C nucleus. Many such parameters are given in tables that appear in Silverstein, Bassler and Morrill.

Become familiar with the chemical shift ranges expected for key functional groups. Some examples are given next page.

^{13}C NMR: Chemical Shifts

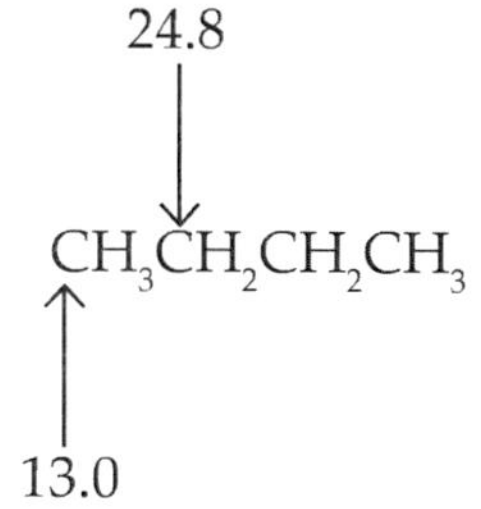

24.8

↓

$CH_3CH_2CH_2CH_3$

↑

13.0

26.9 ↑ (cyclohexane)

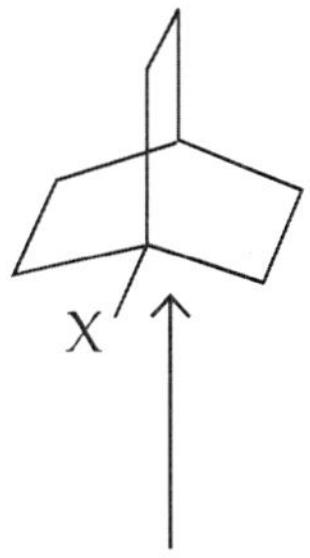

X (bicyclo structure) ↑

X	Chemical Shift (δ)
H	23.9
OCH_3	72.3
F	92.4
Cl	66.2
Br	62.8

CH_3Cl	CH_2Cl_2	$CHCl_3$	CCl_4
24.9	54.0	77.5	96.5

127.4 (cyclohexene)

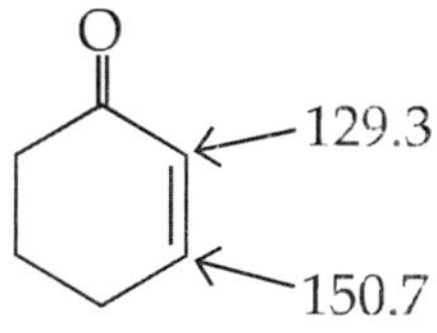

129.3
150.7 (cyclohexenone)

CH_2 ←— 123.3
CH_2

CH_3O—CH ←— 153.2
CH_2 ←— 84.1

CH_3—C(=O)—CH_3 204.1

CH_3—C(=O)—H   200.4

CH_3—C(=O)—OH   178.1

CH_3—C(=O)—OMe   170.7

CH_3—C(=O)—Cl   169.9

CH_3—C(=O)—NMe_2 169.6

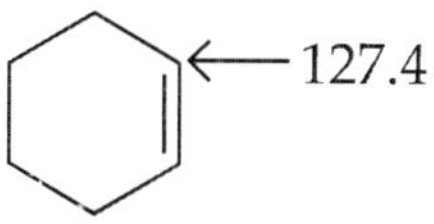

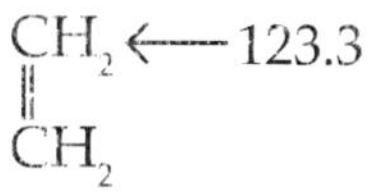

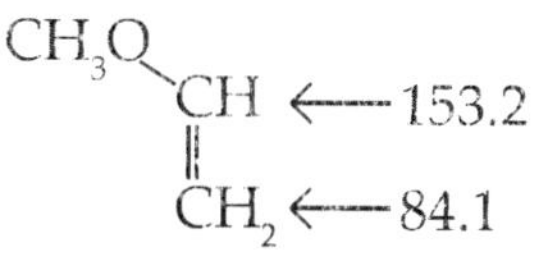

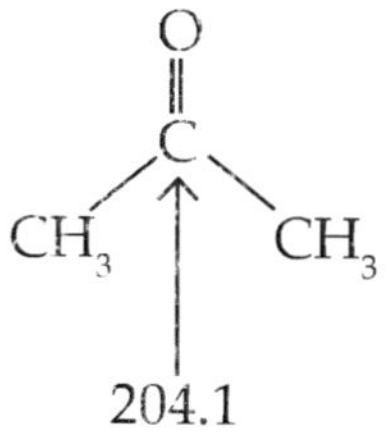

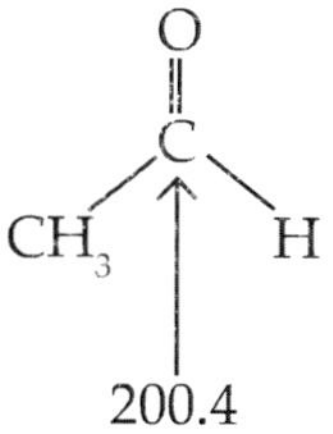

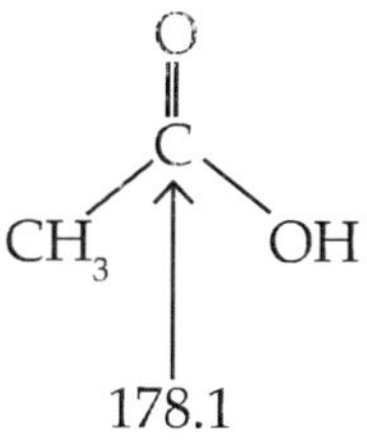

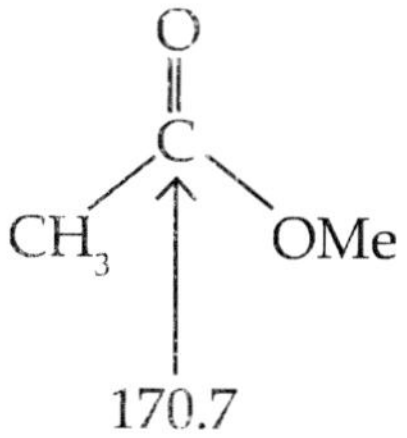

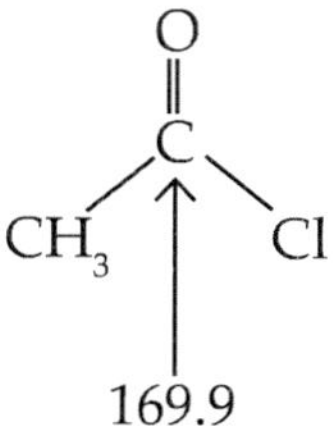

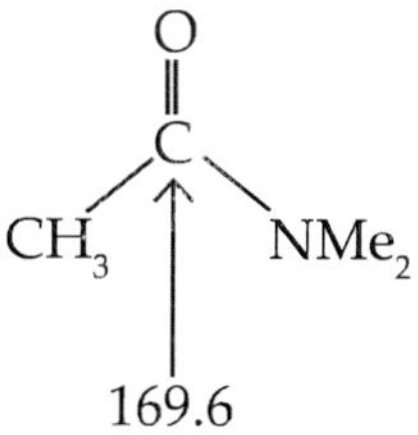

Example

¹H – Decoupled ¹³C NMR Spectra.

Although the ¹H – coupled spectrum of diethyl phthalate is interpretable, more complicated ¹³C spectra are difficult to interpret due to overlap of signals. This problem can be overcome by taking a ¹H – decoupled spectrum. In this experiment a "broad-band" of Rf's that selectively excite the protons are applied continuously while observing the ¹³C's. This procedure affords what is called a "proton noise decoupled" or "broad-band decoupled" or simply a "¹H – decoupled" spectrum. Each carbon appears as a single peak with a characteristic chemical shift.

Note that the peak intensities are still not proportional to the number of carbons. In pulse NMR (FT – NMR), nuclei that "relax" slowly may not be restored to equilibrium before application of second pulse. Since the "relaxation time" of a carbon nucleus is related to the number of protons to which it is bonded (the more protons, the faster the relaxation time), quaternary carbons usually give small signals. In ¹H – decoupled ¹³C spectra, carbons bonded to protons have enhanced signals due to a nuclear Overhauser effect (nOe or NOE). This also accounts for the diminished intensity of quarternary carbon signals relative to CH's, CH₂'s and CH₃'s.

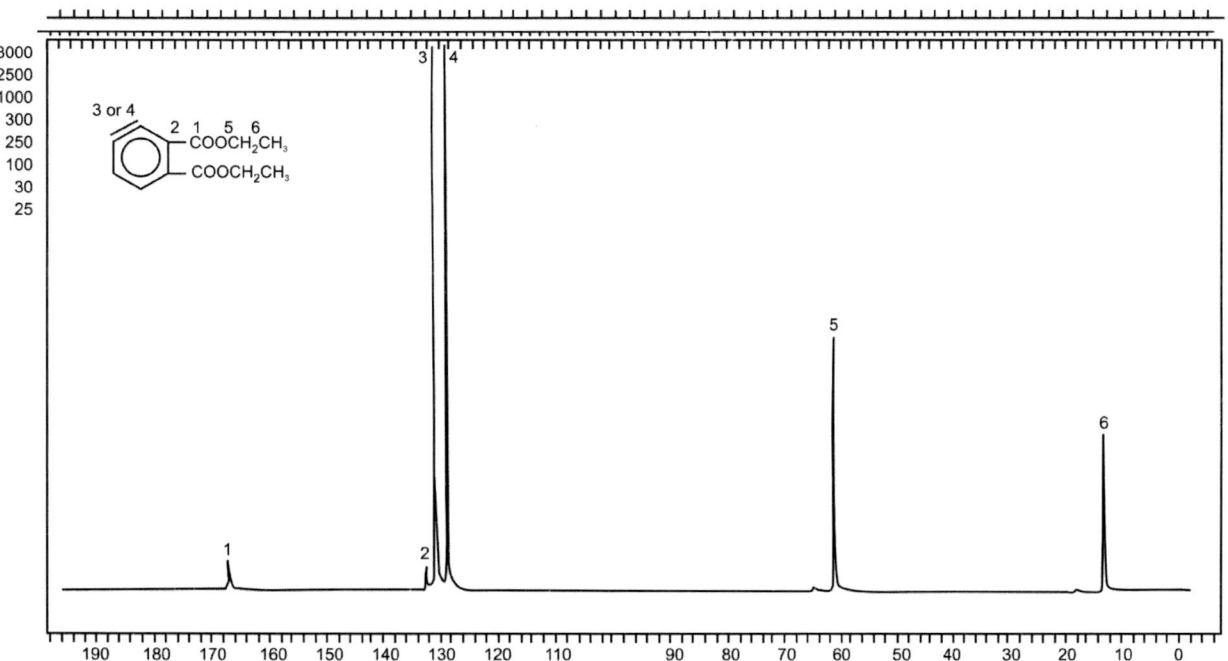

¹³C NMR spectrum of diethyl phthalate with the protons completely decoupled by broad-bond noise. CDCl₃ solvent, 25.2 MHz.

The intensity of the signals due to quaternary carbons can be enhanced by waiting a longer time between pulses (longer "pulse delay"). Notice that although the intensities of the quarterary carbons have increased, the intensities of all signals are still not proportional to the numbers of equivalent carbons. This is due to a difference in the NOE at each ¹³C.

Enhanced Spectrum

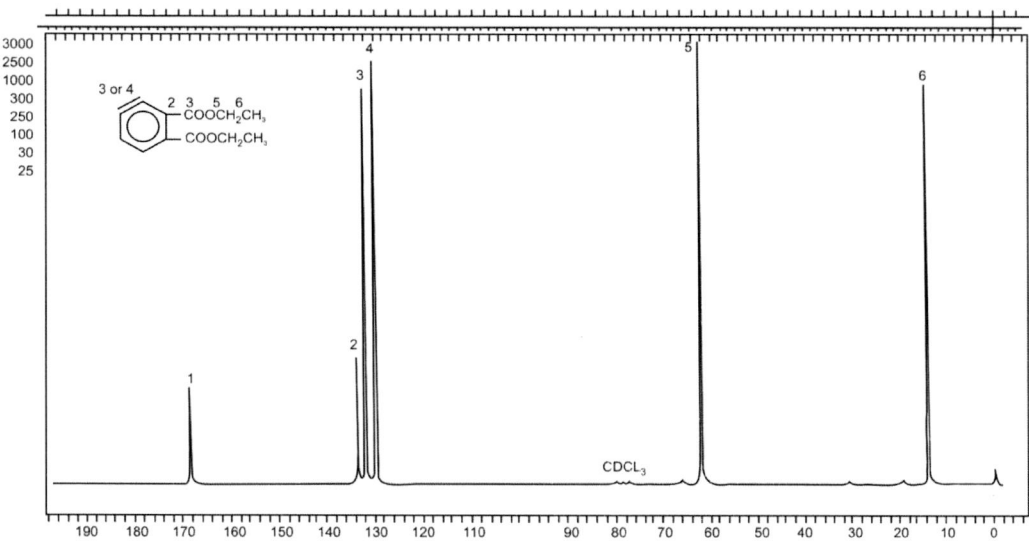

13CNMR spectrum of diethyl phthalate with the protons completely decoupled and a 10 sec. delay between pulses. CDCl₃ solvent, 25.2 MHz.

Off – Resonance Decoupled ¹³C – NMR

While ¹H – decoupled ¹³C spectra provide useful information (number and type of carbons) about the number of hydrogen on each carbon is lost. This information can be obtained by interpreting the frequently complicated ¹H – coupled spectrum or by obtaining an "off-resonance" ¹H – decoupled spectrum. In this experiment, a "broad-band" of decoupling frequencies is applied either upfield or downfield of TMS. The result is a ¹³C spectrum in which some residual ¹³C – ¹H coupling remains and the number of protons on a given carbon can easily be determined. It is important to remember that [symbol here] in such spectra does not represent the ¹H – ¹³C coupling constant.

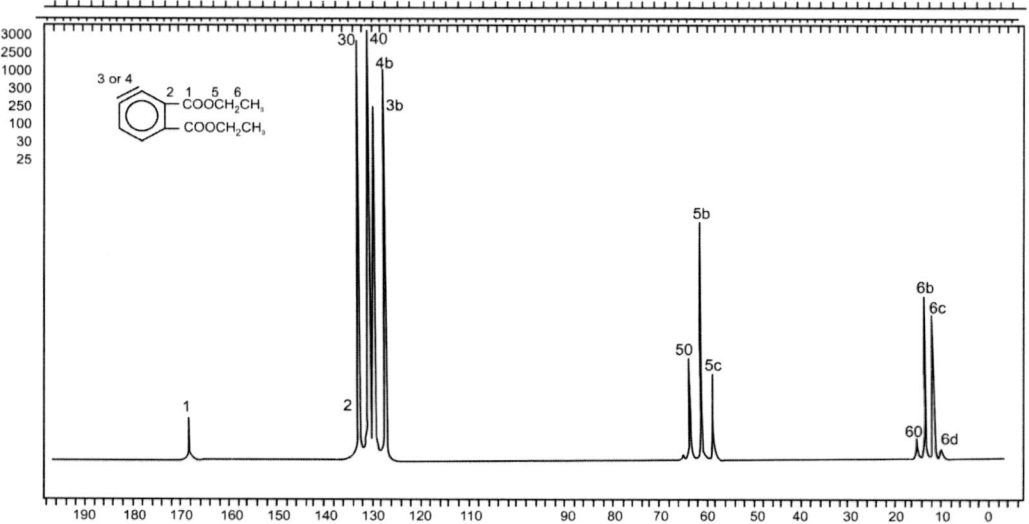

13CNMR spectrum of diethyl phthalate with off-reasonance decoupled protons. CDCl₃ solvent, 25.2 MHz.

Some Structure Determination Problems

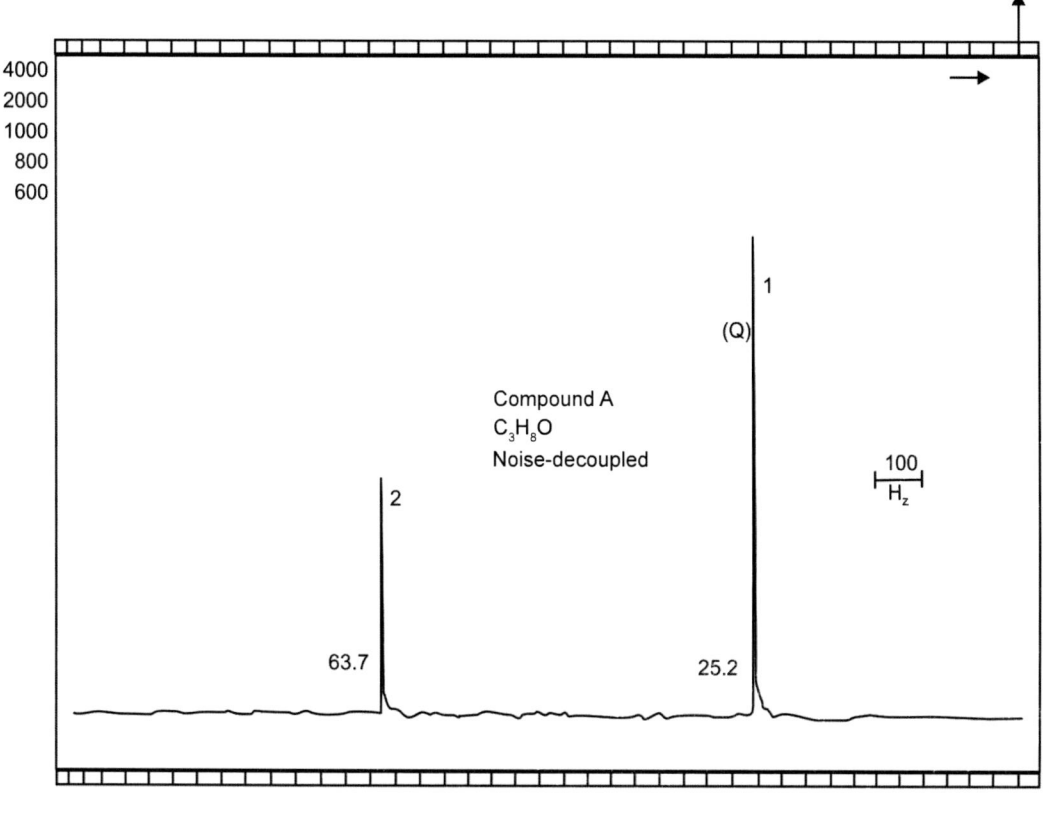

Compound A
C_3H_8O
Noise-decoupled

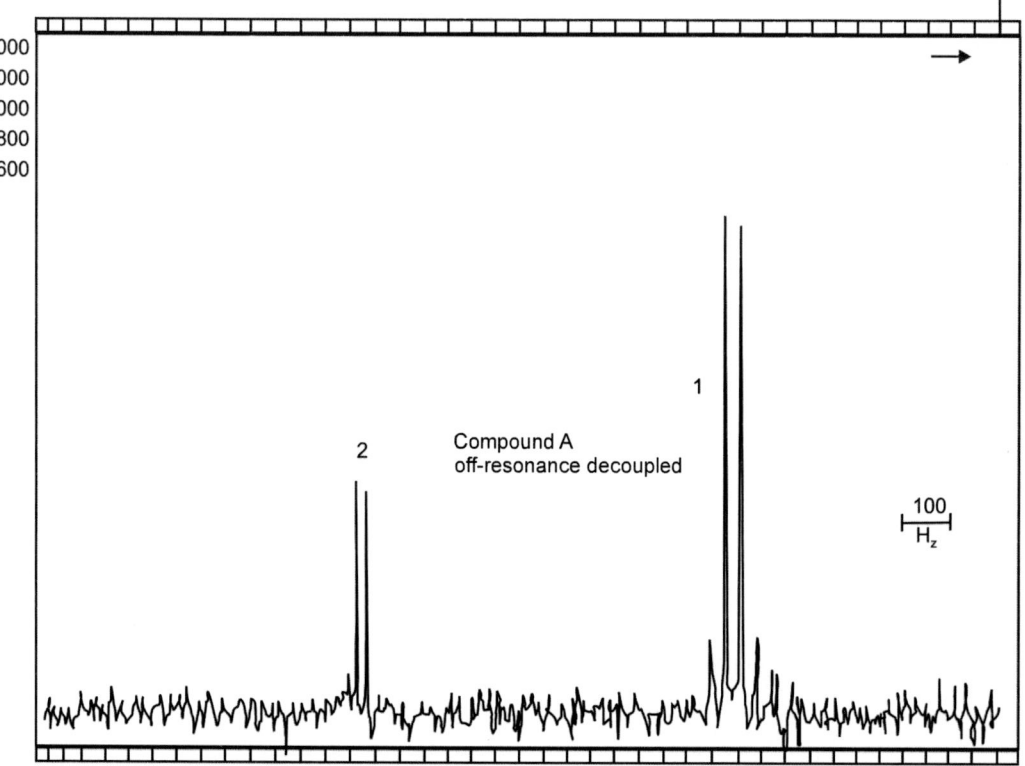

Compound A
off-resonance decoupled

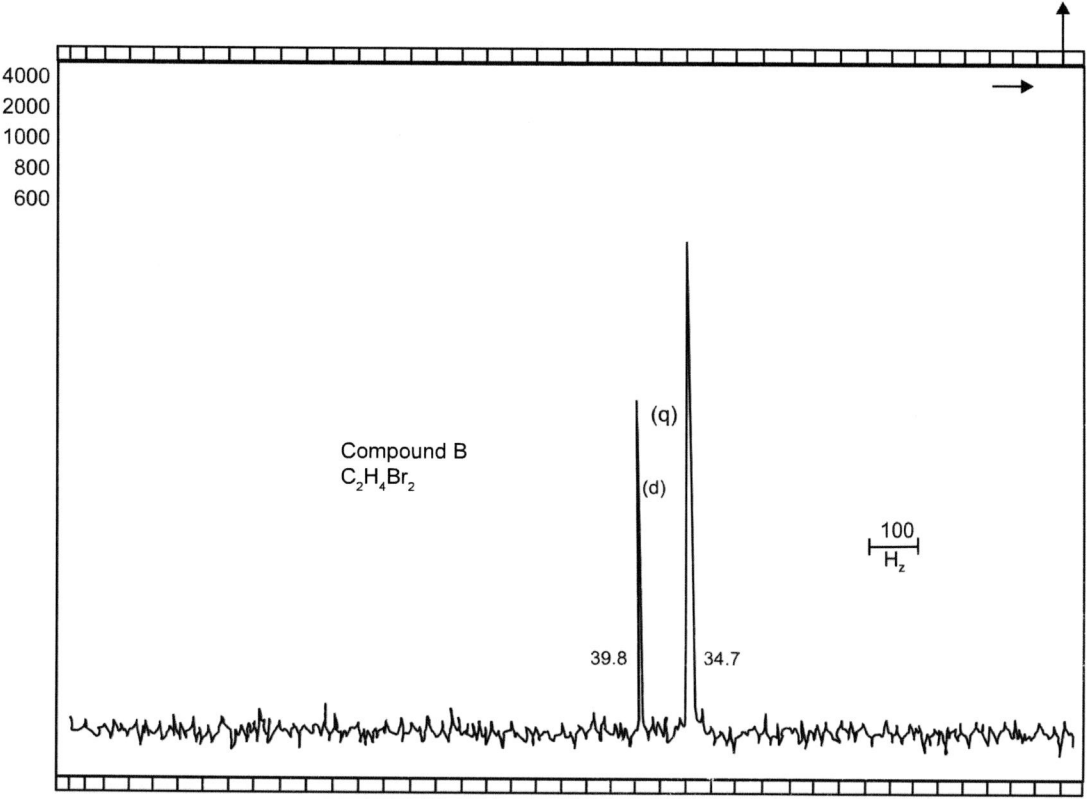

Compound B
C₂H₄Br₂

(q)

(d)

39.8   34.7

100
Hz

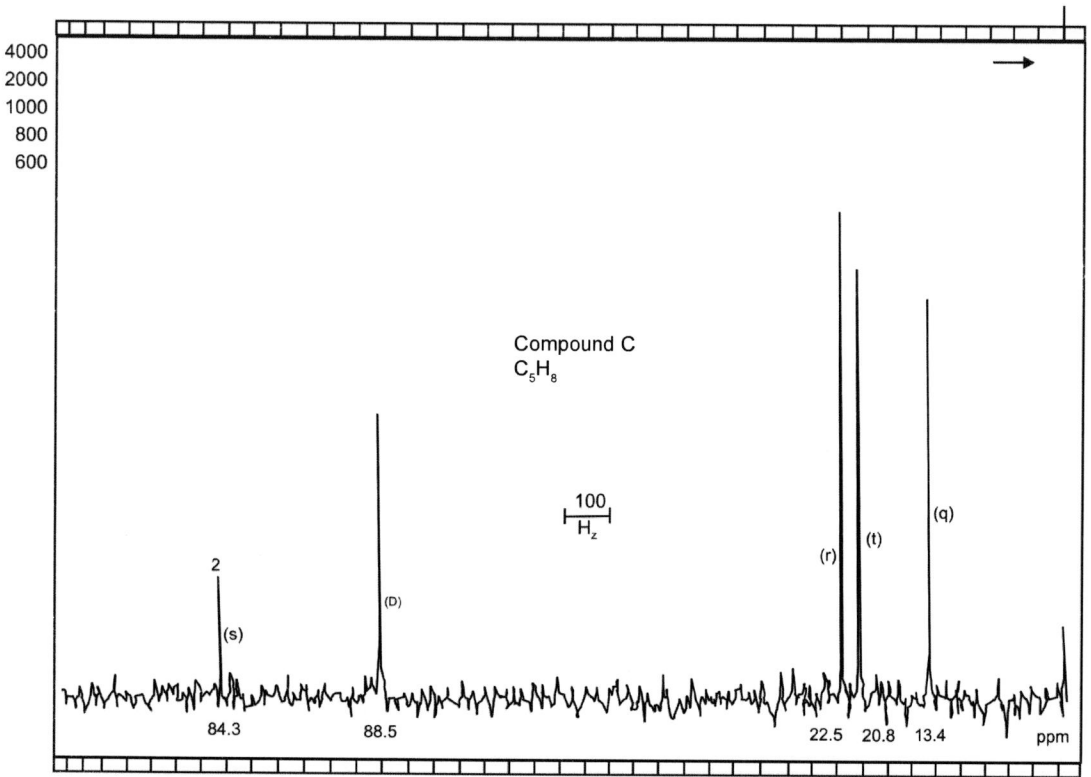

Compound C
C₅H₈

100
Hz

(r) (t)

(q)

2

(s)

(D)

84.3   88.5   22.5 20.8 13.4   ppm

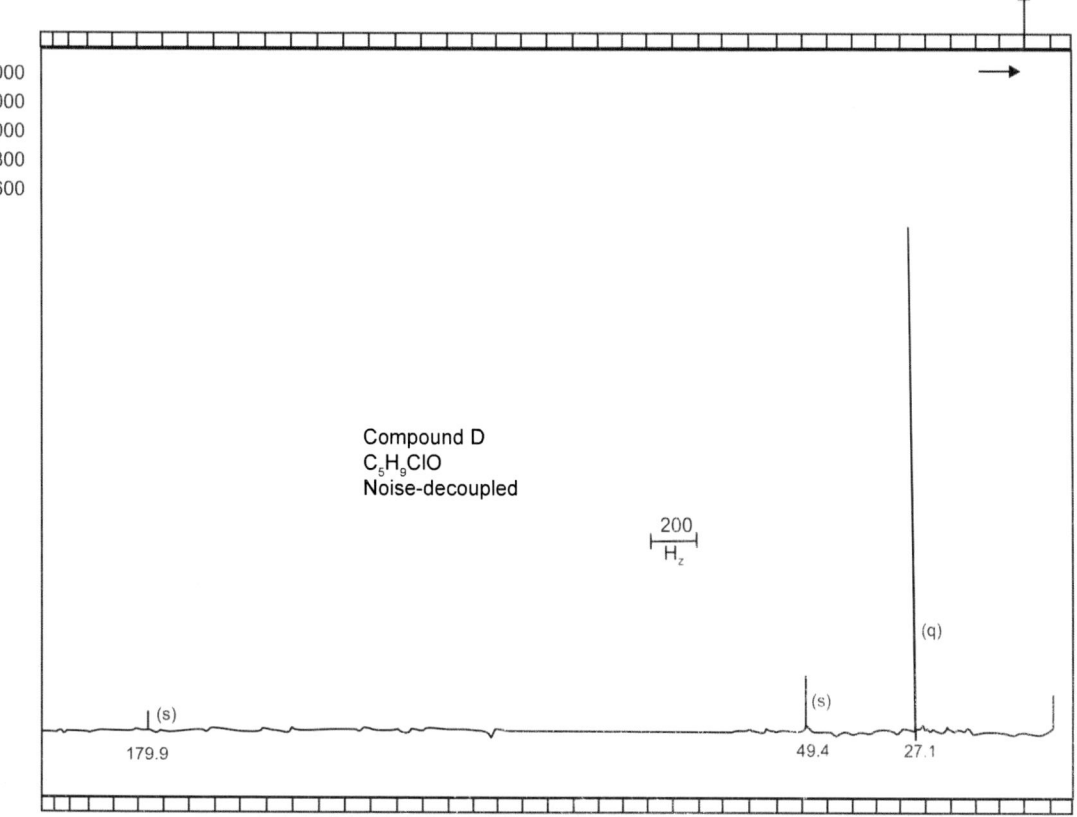

Compound D
C_5H_9ClO
Noise-decoupled

$\overset{200}{\underset{H_z}{\vdash\!\!\dashv}}$

(q)

(s)
179.9

(s)
49.4   27.1

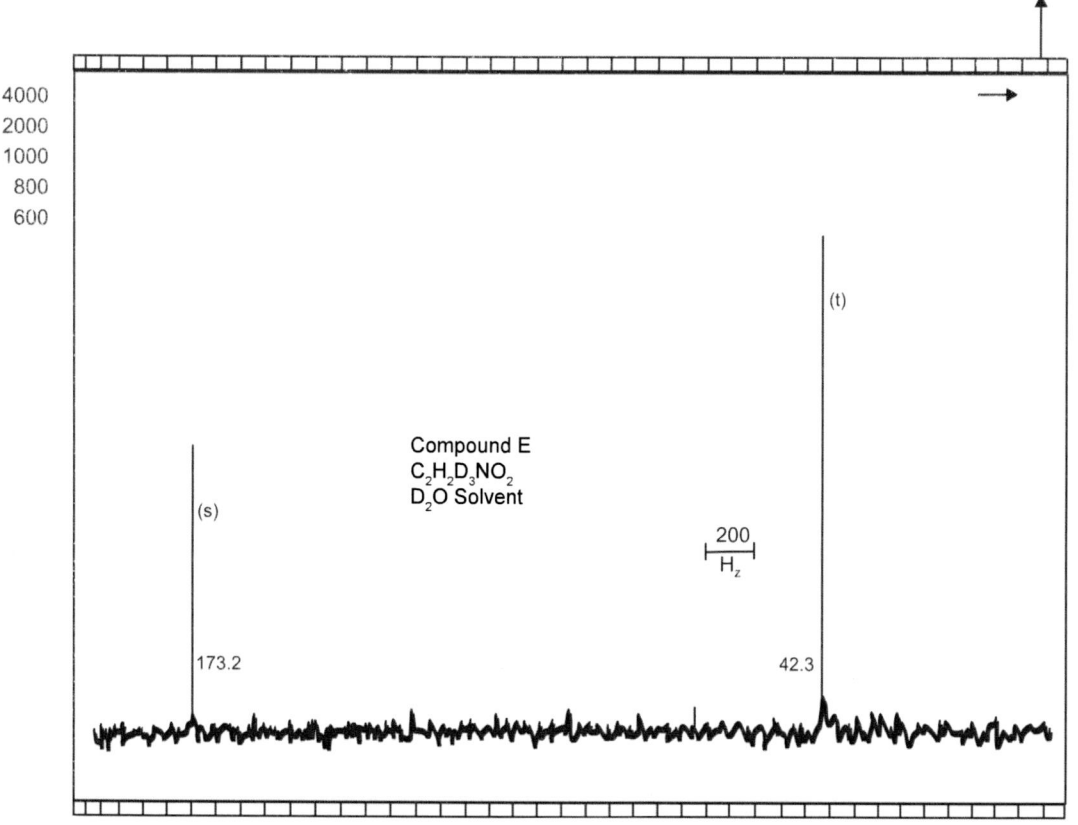

Compound E
$C_2H_2D_3NO_2$
D_2O Solvent

$\overset{200}{\underset{H_z}{\vdash\!\!\dashv}}$

(t)

(s)

173.2   42.3

Distortionless Enhancement by Polarization Transfer (DEPT)

Due to the problems associated with interpreting off-resonance ^1H – decoupled ^{13}C NMR spectra, other techniques have been developed for determining how many hydrogen reside on a given carbon atom. One of the most useful methods involves the DEPT technique. A detailed discussion of this topic is beyond the scope of this course.

DEPT Spectrum

DEPT-NMR spectra for 6-methyl-5-hepten-2-ol. Part (a) is an ordinary broadband decoupled spectrum, which shows signals for all eight carbons. Part (b) is a DEPT-90 spectrum, which shows only signals for the two CH carbons. Part (c) is a DEPT-135 spectrum, which shows positive signals for the two CH and three CH_3 carbons and negative signals for the two CH_2 carbons.

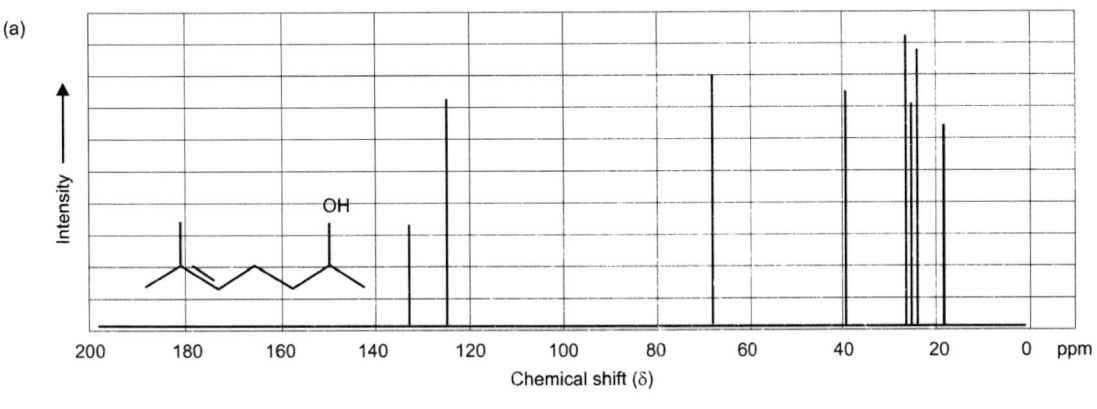

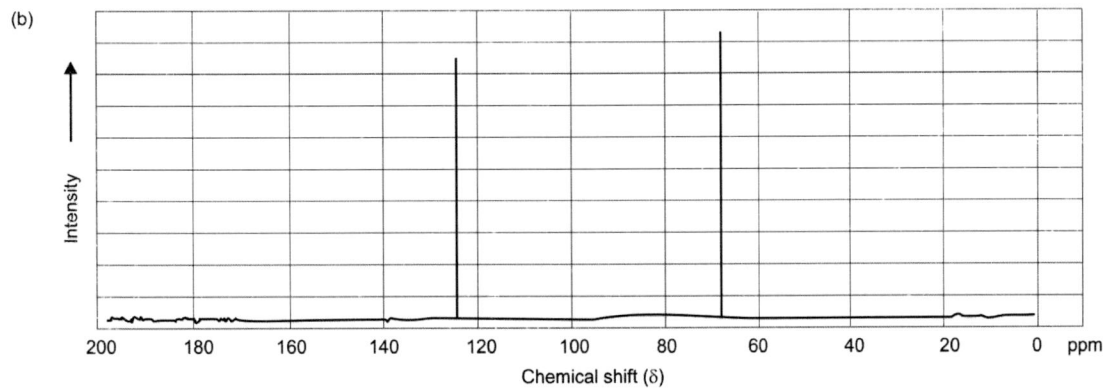

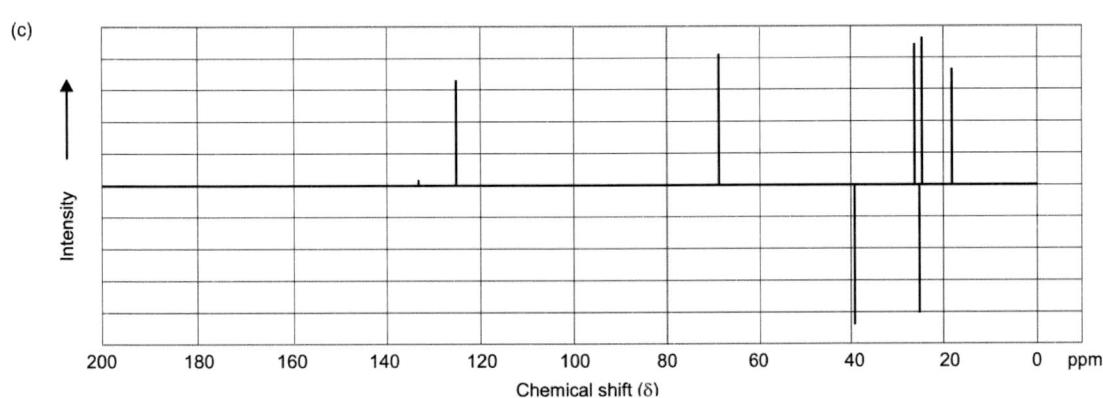

Mass Spectrometry

A mass spectrum is a record of ions (usually positively charged) formed upon exposure of a molecule in the gas phase to an "ionizing environment". The ionization can be accomplished in a number of ways, the classical manner being exposure of the compound to an electron beam.

A typical "electron impact" mass spectrometer consists of the following components:

Sample Handling System: This is used to introduce a neutral molecule to the ionization chamber.

Ionization and Accelerating Chamber: This chamber is usually operated at a pressure of 10^{-5} to 10^{-6} Torr and consists of a beam of high energy electrons (e.g. 70 ev) used to ionize the sample:

$$M \rightarrow M^{+} + \text{-} \, e^{-}$$

and devices that focus and accelerate the resulting ions into an "analyzer". It usually requires at least a 10 ev beam of electrons (1 ev = 23 kcal mol-1) to remove an electron from an organic molecule. When a 70 ev electron beam is used, the additional energy of the electrons may be dissipated in breaking bonds of the molecular ion. This produces smaller ions that can also be accelerated into the "analyzer".

Analyzer Tube: For low resolution instruments (instruments that can distinguish between ions that differ by one mass unit such as m/e 100 and m/e 101), the analyzer tube is a curved magnetic field that deflects the ions according to the equation:

$$m \, / \, e = H^{2} \, r^{2} \, / \, 2V$$

where H is the strength of the magnetic field, r is the radius of the circular path in which the ion is traveling and V is the accelerating potential.

Schematic of a Single-Focusing Mass Spectrometer

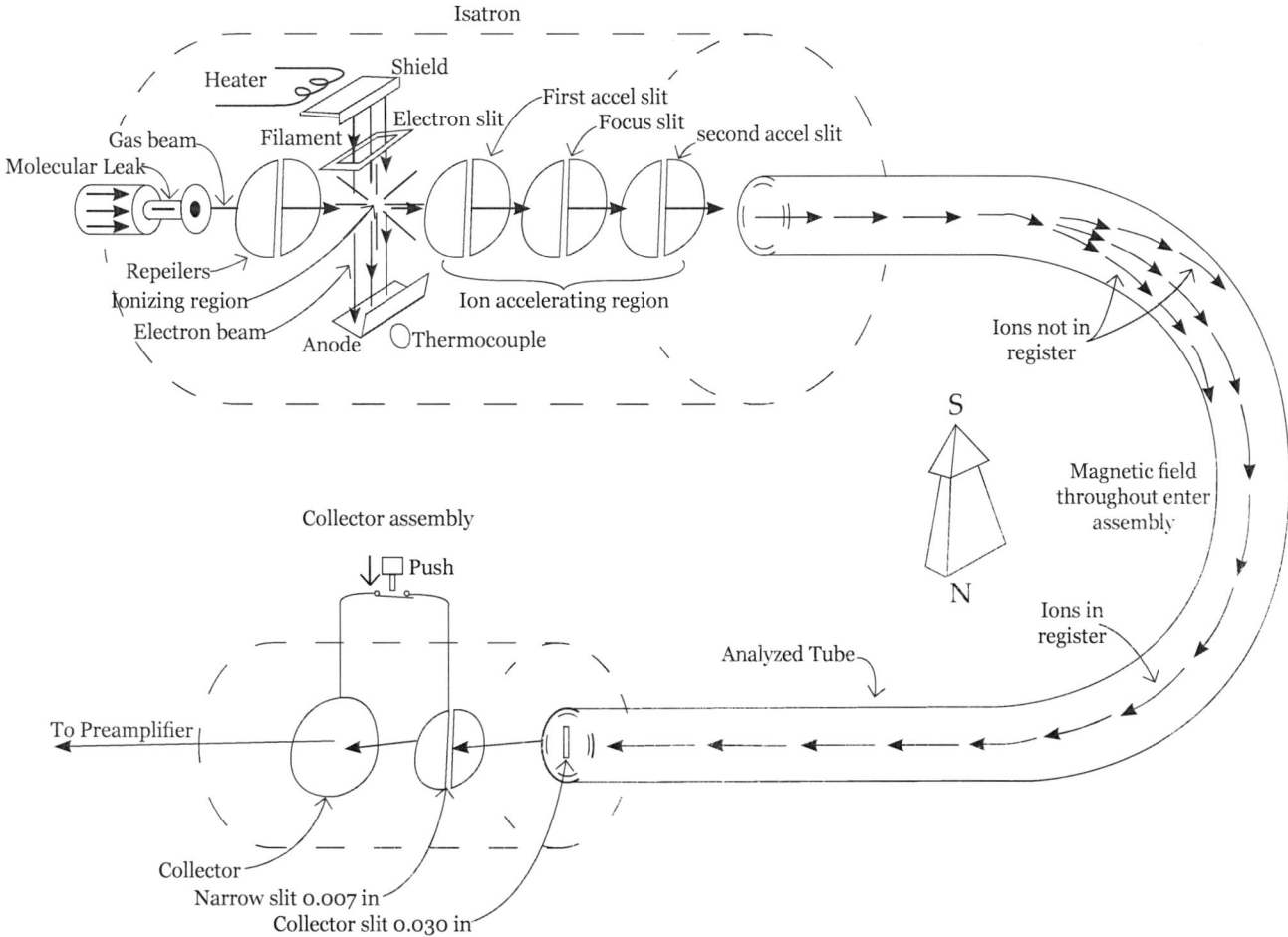

Schematic diagram of CEC model 21-103 Mass Spectrometer. A single-focusing. 1800 sector mass analyzer. The magnetic fields is perpendicular to the pare.

Double-Focusing High Resolution Mass Spectrometers

Sometimes one wants to differentiate between ions which have the same nominal mass. For example carbon monoxide (CO), ethane ($H_2C=CH_2$) and nitrogen (N_2) all have a nominal mass of 28. Their actual masses, however, are 27.9949, 28.0313, and 28.0061. These masses can be separated using a "high resolution" mass spectrometer. Such instruments consist of an electrostatic analyzer in conjunction with the magnetic analyzer. The electrostatic analyzer is used to "focus" ions of equal energy into the magnetic analyzer. For any given m/e value, mono energetic ions can be accurately focused at the collector and ions whose masses differ by little as a few parts per million can be separated.

Element	Atomic Weight	Nuclide	Mass
Hydrogen	1.00797	^{1}H	1.00783
		D.(^{2}H)	2.01410
Carbon	12.01115	^{12}C	12.000000 (std)
		^{13}C	13.00336
Nitrogen	14.0067	^{14}N	14.0031
		^{15}N	15.0001
Oxygen	15.9994	^{16}O	15.9949
		^{17}O	16.9991
		^{18}O	17.9992
Fluorine	18.9984	^{19}F	18.9984
Silicon	28.086	^{28}Si	27.9769
		^{29}Si	28.9765
		^{30}Si	29.9738
Phosphorus	30.974	^{31}P	30.9738
Sulfur	32.064	^{32}S	31.9721
		^{33}S	32.9715
		^{34}S	33.9679
Chlorine	35.453	^{35}Cl	34.9689
		^{37}Cl	36.9659
Bromine	79.909	^{79}Br	78.9183
		^{81}Br	80.9163
Iodine	126.904	^{127}I	126.9045

Inlet Systems and GC-MS

Most organic compounds have insufficient thermal stability to be heated to temperatures where usable vapor pressures can be achieved. In such cases a direct-insertion technique is used. A sample is placed on a probe that is inserted directly into the ionization chamber. The low pressures achieved in the chamber and devices that allow one to heat the probe allow one to obtain mass spectra of high molecular weight compounds such peptides and polysaccharides. Another useful method is GC-MS. Such instruments are usually limited to unit mass resolution. A schematic diagram is provided below.

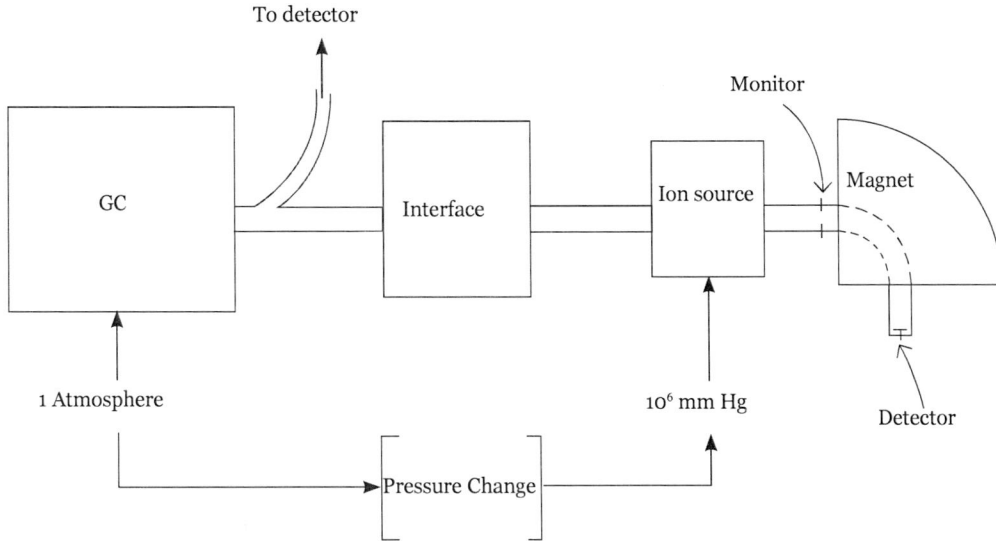

Organic compounds with high volatility can be introduced to the ionization chamber of a mass spectrometer using a device of the type shown below. Various combinations of temperature and pressure can be used to obtain usable gas phase concentrations of molecules.

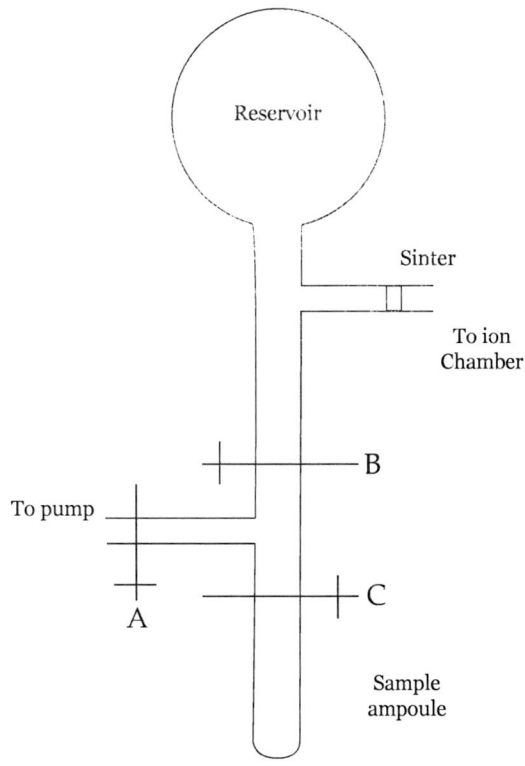

MASS SPECTROMETRY (MS)

Purpose

- Technique for measuring the mass, (Molecular Weight) of a molecule.

- Measuring the masses of the fragments produced, it is possible to obtain structural information about a molecule.

- Molecular formula could be derived from a molecular weight.

Important Information on MS

1. **Base Peak:** The most abundant ion formed in the ionization chamber give rise to the tallest peak in the mass spectrum.

2. **Molecular ion (M⁺):** Molecular Weight of the molecule – The simple removal of an electron from a molecule.

3. **M + 1 and M + 2:** Refers to isotopes – They are one or two mass unit above the mass of the "normal" atom.

4. **Fragment ions:** When some of the bonds are broken in the molecule, series of molecular fragments are produced. These fragments appear as peaks on the spectrum. These fragment ions appear at m/e values corresponding to their individual masses.

Mass Spectra of Ethers

A major fragmentation pathway for ethers is "α –cleavage". Another fragmentation pathway involves O-alkyl cleavage to a carbocation.

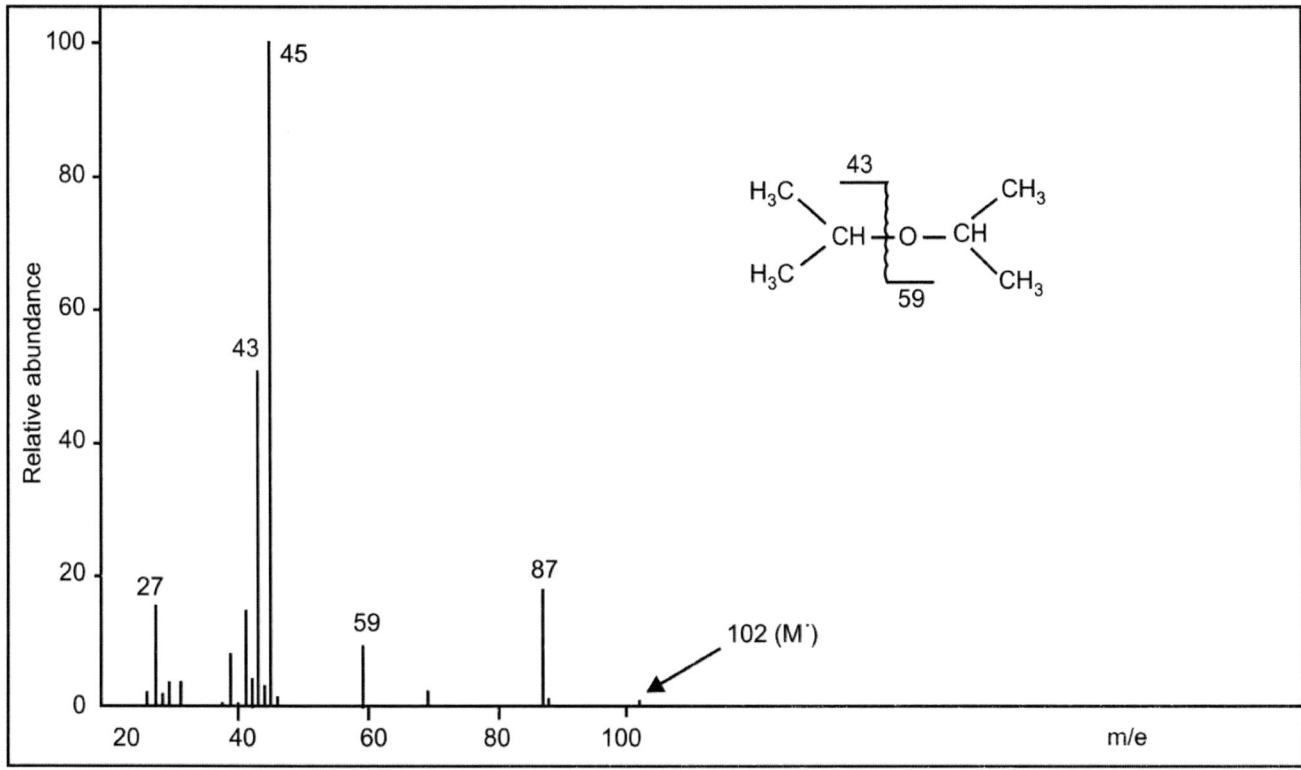

Note that the molecular formula of $C_6H_{14}O$ suggests that the compound is an ether or an alcohol, but the absence of an M-18 peak suggests an ether.

Mass Spectra of Alcohols

Important Features

- α-Cleavage remains an important fragmentation pathway.

- M-1 peaks are frequently large.

- Loss of H_2O is a major fragmentation pathway

- Molecular ions are frequently weak and in the case of tertiary alcohols are often absent.

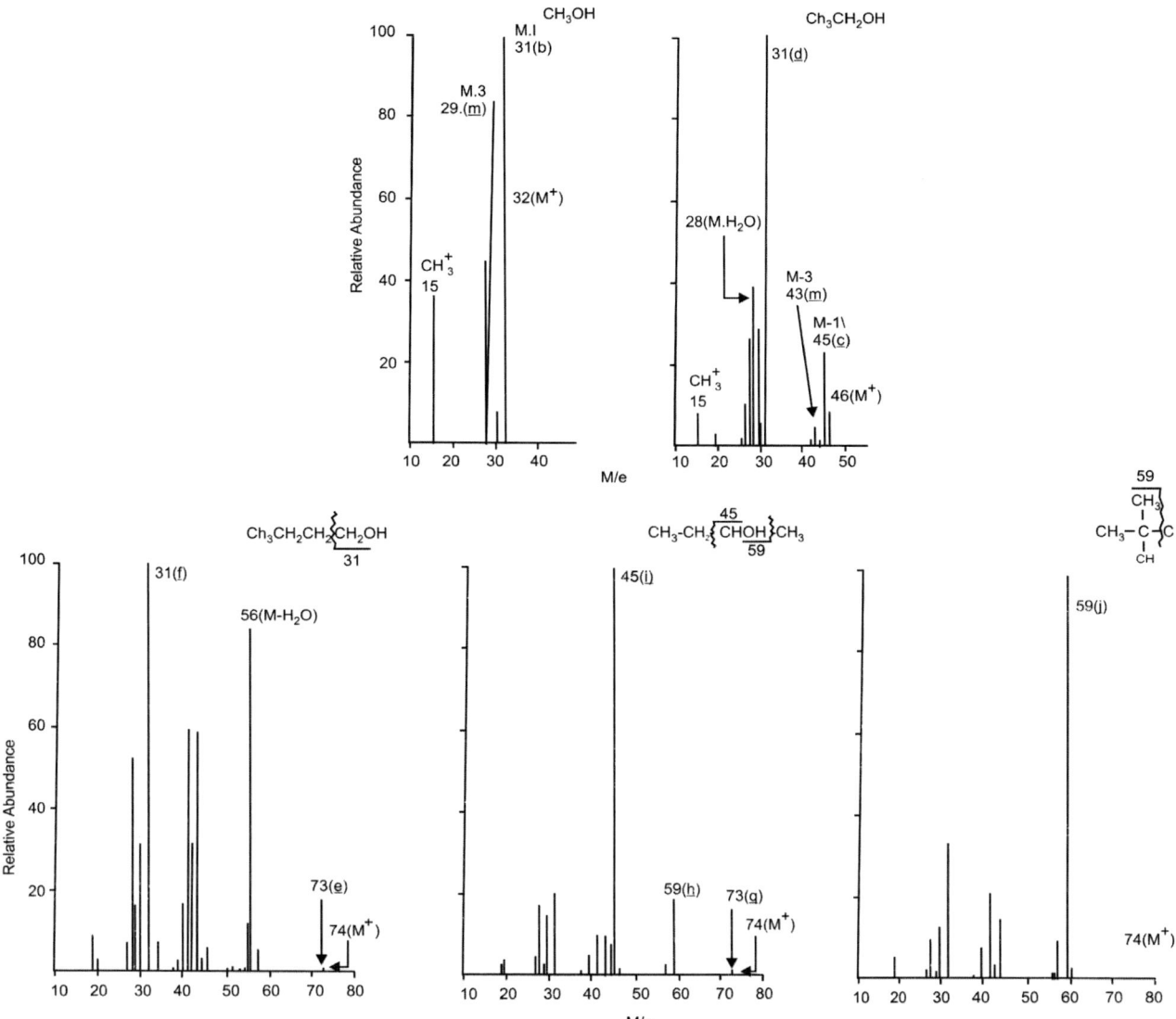

Mass Spectra of Amines

Important Features

- Odd molecular weight and ions at m/e 30, 44, 72, and 114 suggest a saturated amine.

- The largest alkyl radical is preferentially lost in the α-cleavage of the molecular ion.

Mass Spectra of Carbonyl Compounds

Carbonyl compounds give peaks due to α-cleavage. They also give diagnostic peaks due to the McLafferty rearrangement (hydrogen atom is transferred to the carbonyl oxygen, a C-C bond is broken).

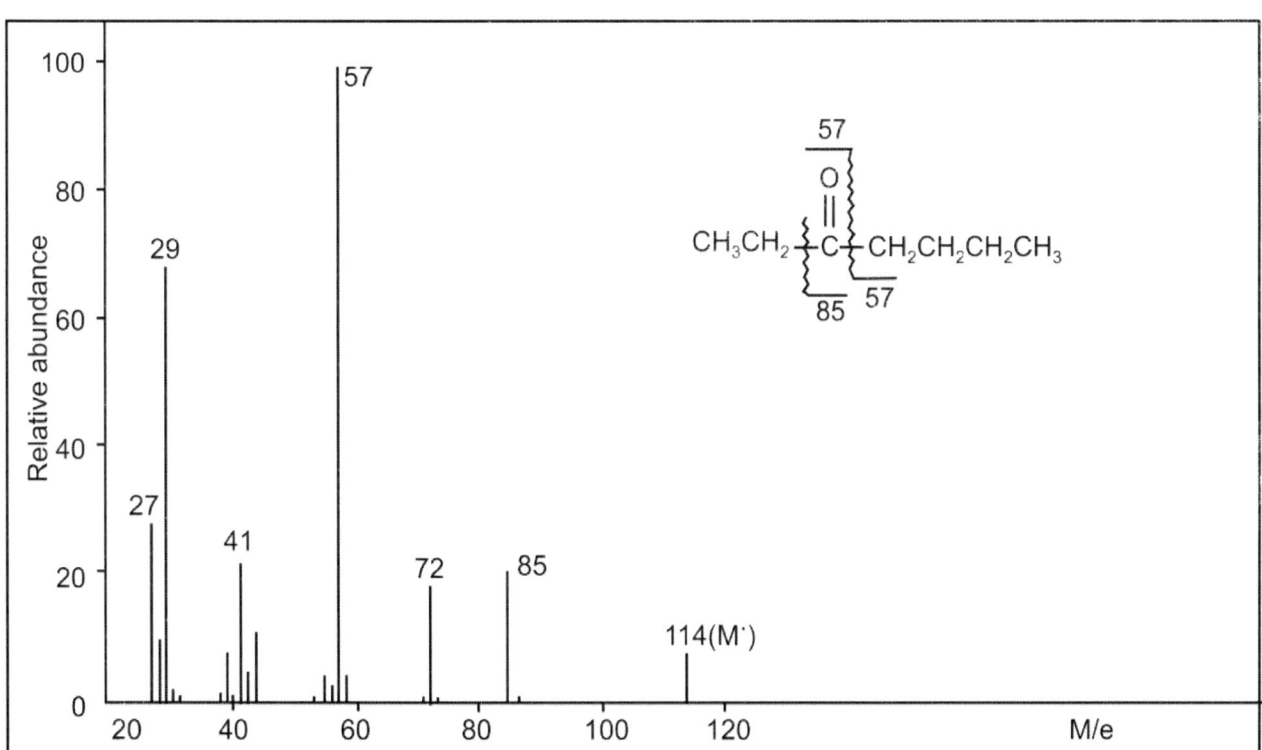

α- Cleavage

Mass Spectrum of an Aldehyde

- M – 1 peak and small peak for loss of water (M-H₂0).

- The peak at m/e 44 is due to the McLafferty rearrangement.

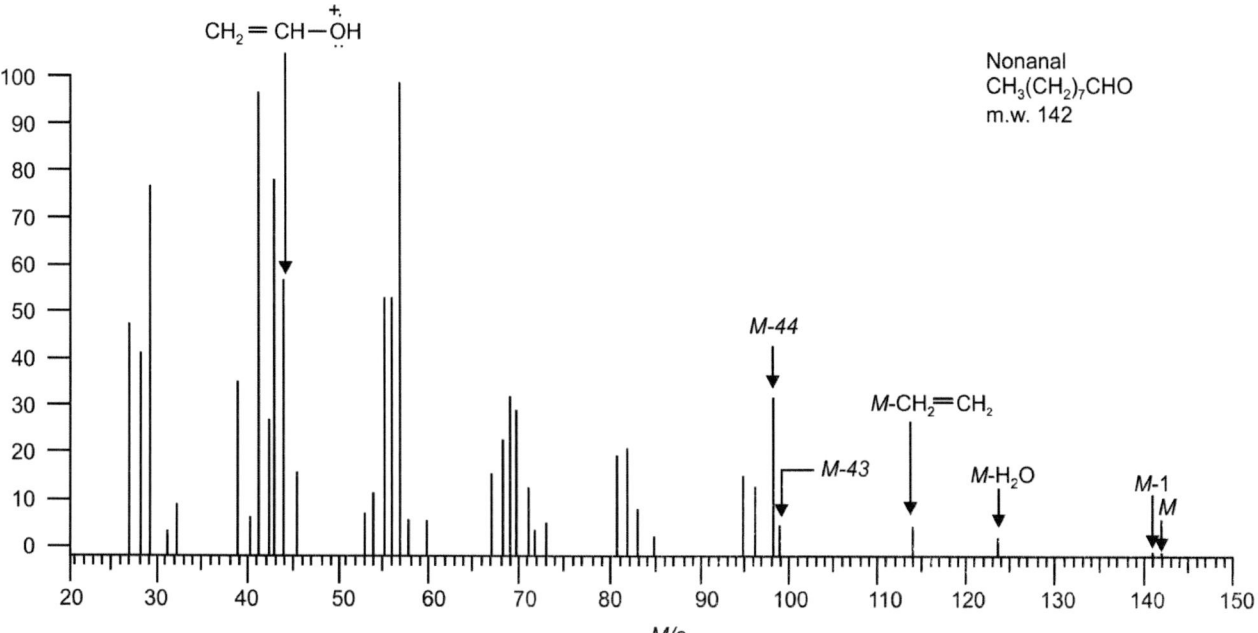

Mass Spectrum of an Ester

- Cleavage on both sides of the carbonyl group.

- The peak at m/e 74 is due to the McLafferty rearrangement.

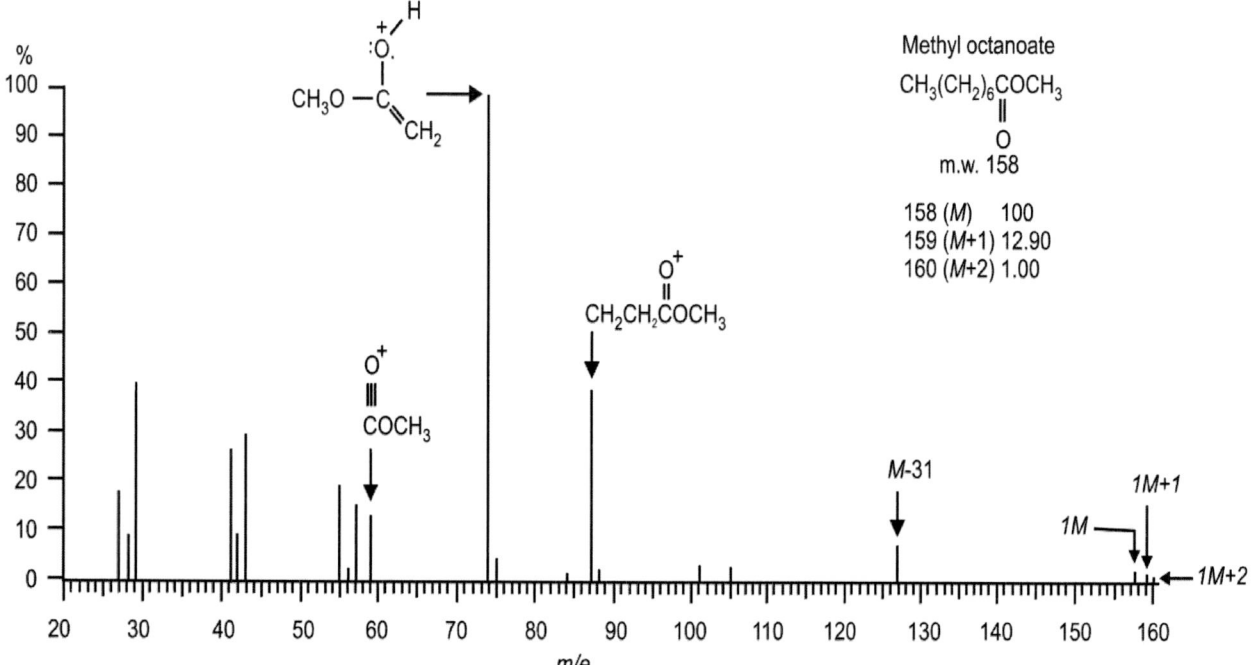

How to use molecular weight to determine a possible molecular formula.

Method: Rule of Thirteen

Formula: $\dfrac{MW}{13} = n + \dfrac{r}{13}$

Explanation: 13 = Mass of 1 carbon and 1 hydrogen

Look at C_nH_{n+r}

Assumption: Hydrocarbon Molecule

Example: A Compound with MW of 200

$$\frac{MW}{13} = n + \frac{r}{13}$$

$n = whole\ number$

$r = MW - (n \times 13)$

$$\frac{200}{13} = 15 + \frac{5}{13}$$

Possible molecular formula $= C_nH_{n+r}$

$$= C_{15}H_{20}$$

Addition of other atoms

For oxigeon

-     Subtract an equivalent of atomic mass from the hydrocarbon molecular formula

$$C_{15}H_{20} + O\text{-}CH_4 = C_{14}H_{16}O$$

Degree of unsaturation (DU) Hydrogen deficiency index (HD)

-     Helps to determine the number of unsaturation in a molecule.

-     Number of unsaturation will help in the drawing of a possible structure.

How to determine unsaturation

$$DU = \frac{Number\ of\ H_{if\ sat.} - Actual\ Number\ of\ H}{2}$$

$saturated = C_nH_{2n}$

$unsaturated = C_nH_{2n} = C_nH_{2n-2}$

Information from DU

- Double bounds=1 unsaturation

- Triple bonds = 2 unsaturation

- Ring = 1 unsaturation

Example: $C_{15}H_{20}$

$$DU = \frac{32 - 20}{2} = 6$$

Example 2: C_6H_6O

a. Determine DU

b. Draw 2 possible structures

solution

a. $DU = \dfrac{14 - 6}{2} = 4$

b.

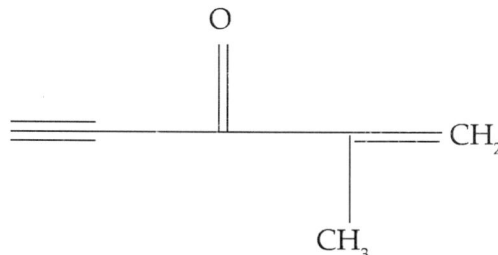

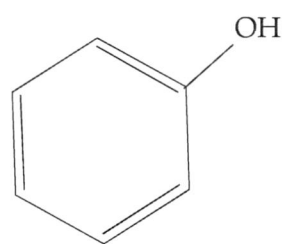

Note:

For benzene ring (Aromatic)

- DU > or = 4

- # C > or = 6

For Compounds containing nitrogen or halogens, The DU is:

$$DU = \frac{\text{\# of H}_{\text{if sat}} - \text{Actual \#H} + \text{\#N} - \text{\# Halogen}}{2}$$

*   Odd molecular weight implies odd number of nitrogen.

Example 3: C$_8$H$_9$N

$$DU = \frac{18 - 9 + 1}{2} = 5$$

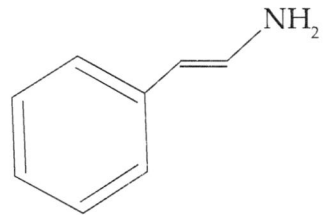

Determine DU and draw a possible structure.

Example 4: C$_6$H$_5$Br

SPECTRAL PROBLEMS

In this section, you will test your understanding of all the spectroscopic methods that you have covered in this course. The following problems will give you the opportunity to apply what you have learned in this course. The problems include the infrared (IR) spectrum, the mass spectrum (MS), and carbon and proton (^{13}C and ^1H) NMR spectral data. Molecular formula for some of the compounds has been provided.

These problems could be solved using different strategies. In general, always start with the spectrum you understand best. Below are some helpful hints:

1. **Start with infrared (IR) spectrum**

   Identify all visible functional groups
   Write down the functional groups and their frequencies

2. **Proton NMR Spectrum**

   Determine the number of different hydrogen chemical environments
   Examine splitting patterns for information about the number of nearest nonequivalent hydrogen neighbors.
   Write down the structural fragments. CH_3, CH_3CH_2, CH_3CHCH_3, CH_2, aromatic protons (mono-substituted, para-substituted, etc...).

3. **Carbon NMR Spectrum**

   Determine the number of carbons in different environments and the number of methyl carbons, methylene carbons, methane carbons, and quaternary carbons.

4. **Mass Spectrum**

   Determine the molecular weight from the spectrum (last peak on spectrum) if not given.
   Determine the molecular formula using the "Rule of 13" or any other method.
   Use functional group(s) to adjust the molecular formula.
   Odd molecular weight implies odd number of nitrogen.
   Calculate Degree of unsaturation (DU) or Hydrogen deficiency index (HDI).

5. **Draw** possible structures keeping in mind that there may be more than one way putting the fragments together. Remember ISOMERS? That's exactly what you are doing at this point. The proposed structures should match the spectra analyses.

COMBINED IR, MS, ^{13}CNMR, and ^{1}HNMR PROBLEMS

Determine the molecular structure for a molecule with mass of 198. Assume that C, H, and Br may be present. The NMR and IR spectra of the compound is also provided to aid you in your decision. Show all calculations to receive full credit.

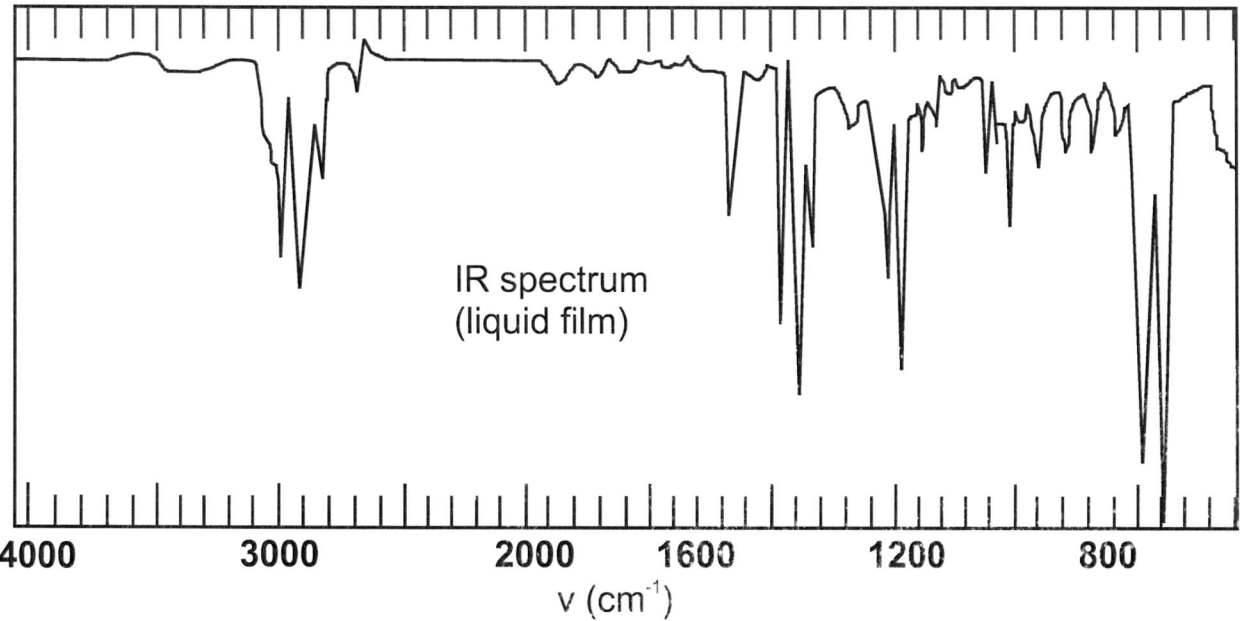

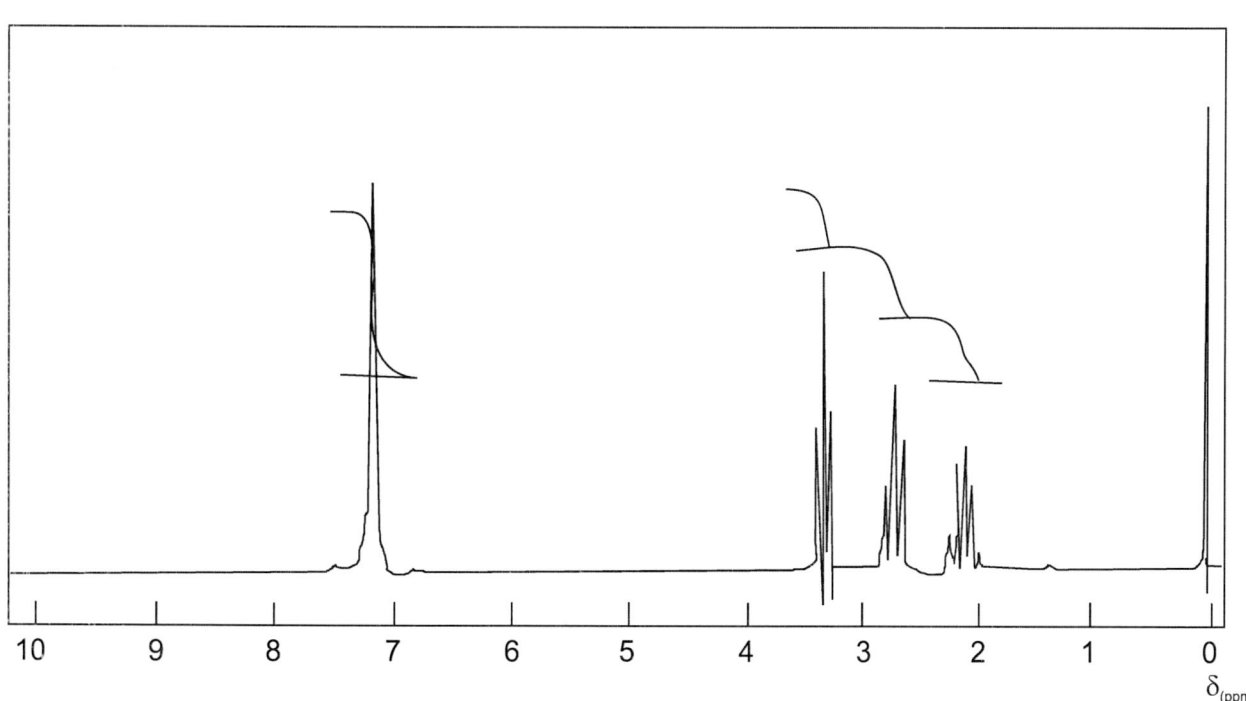

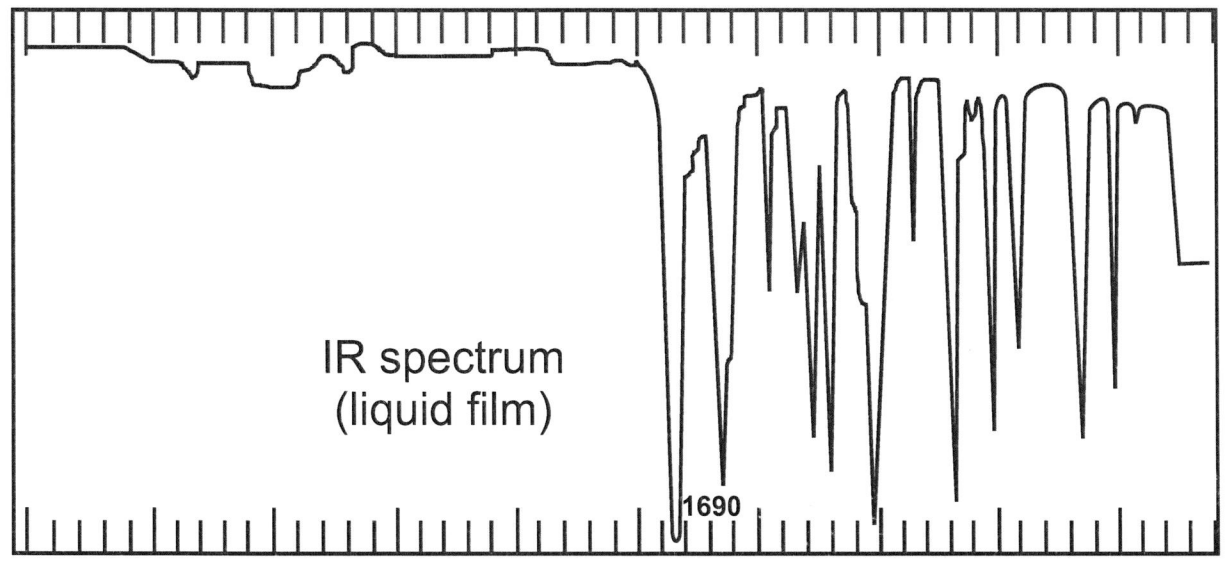

IR spectrum
(liquid film)

1690

V (cm⁻¹)

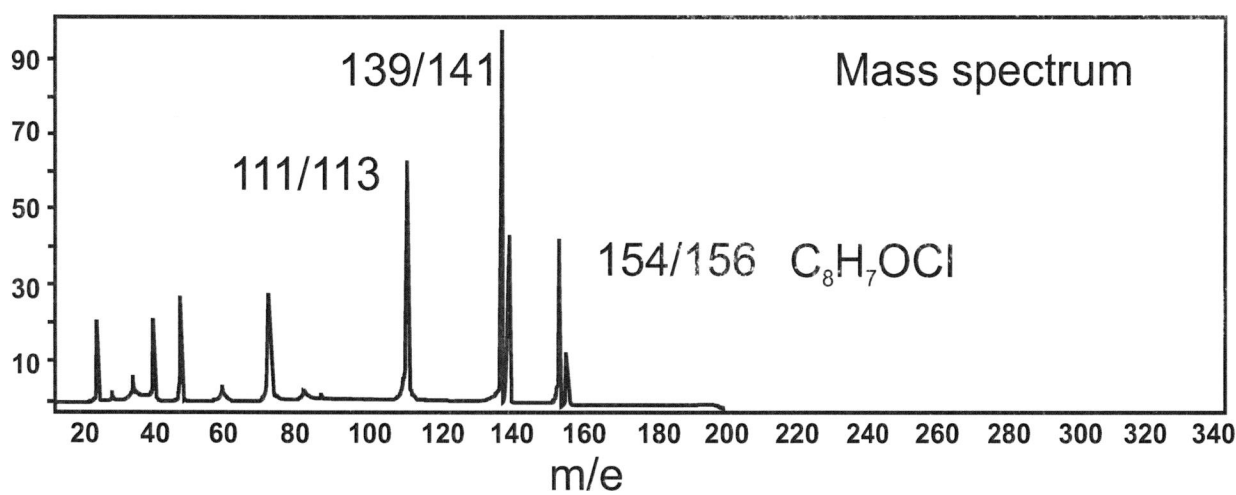

Mass spectrum

139/141

111/113

154/156 C₈H₇OCl

m/e

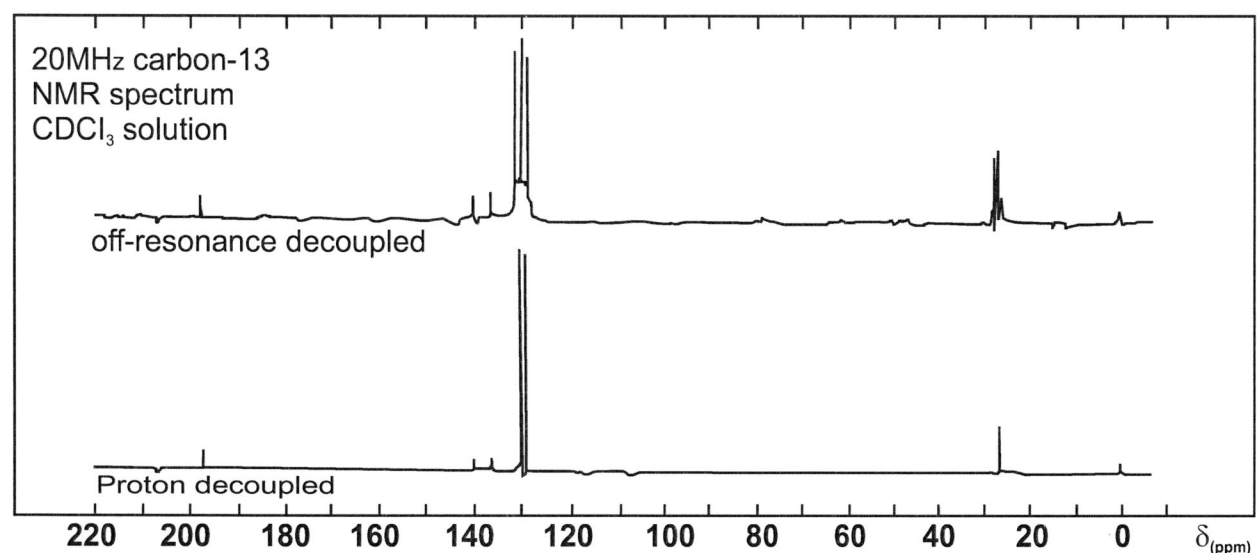

20MHz carbon-13
NMR spectrum
CDCl₃ solution

off-resonance decoupled

Proton decoupled

$\delta_{(ppm)}$

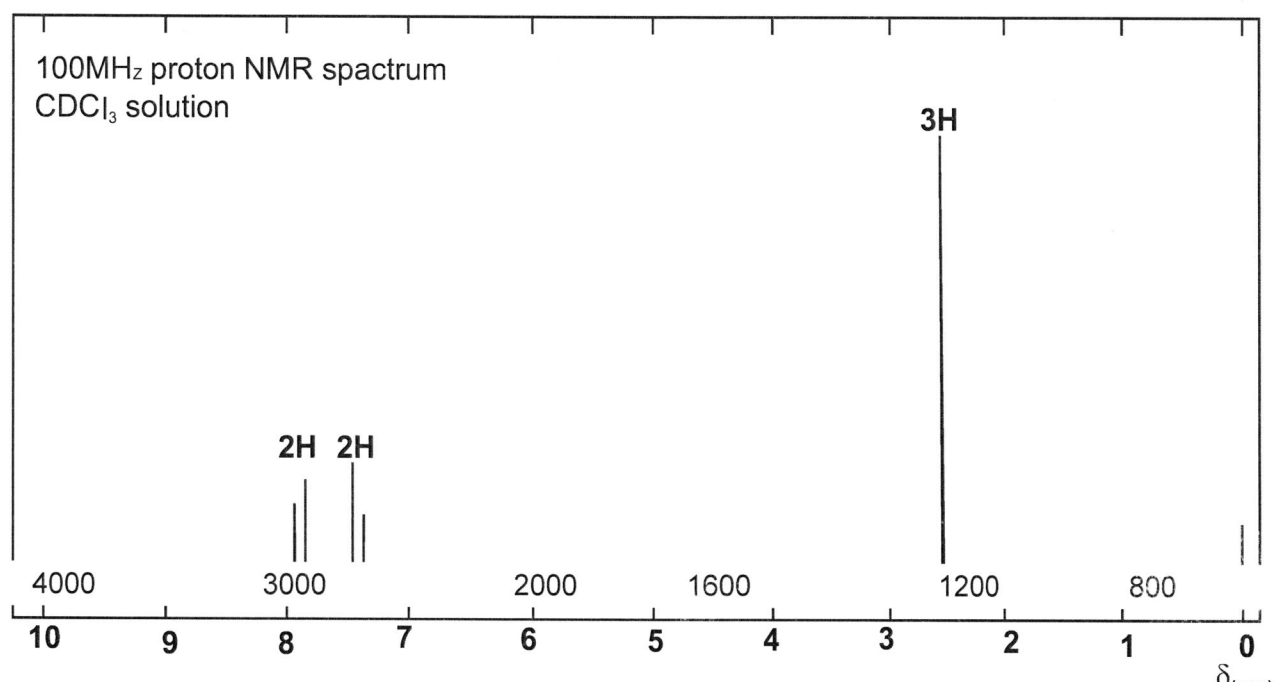

100MHz proton NMR spactrum
CDCl₃ solution

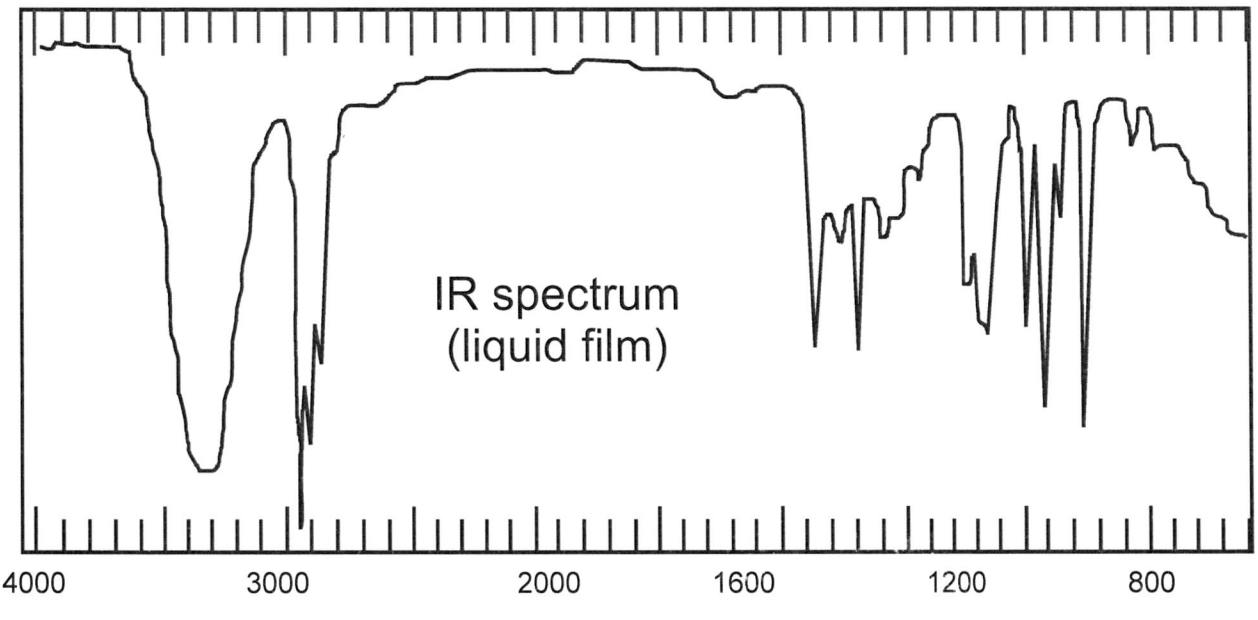

IR spectrum
(liquid film)

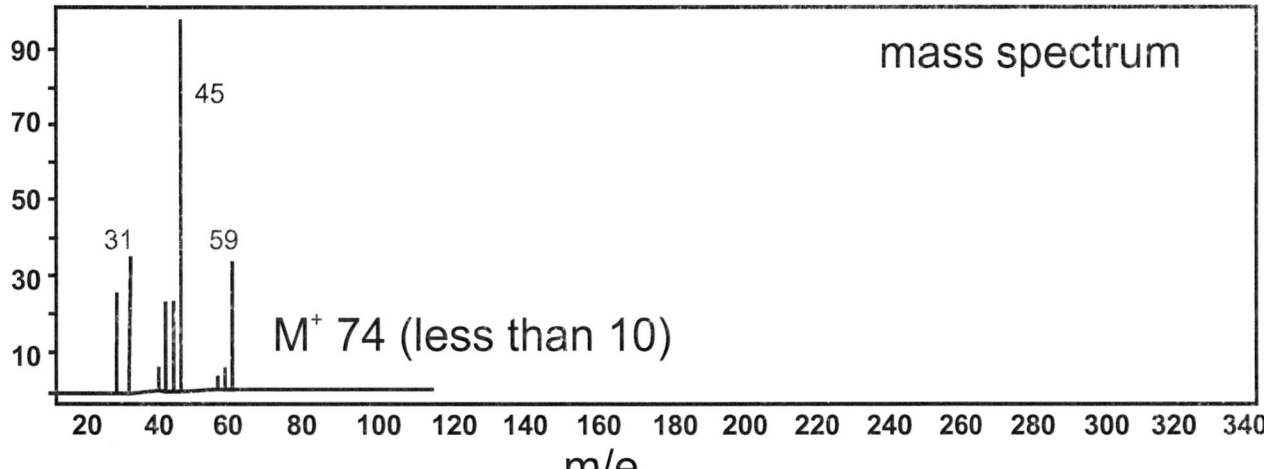

mass spectrum

45

31

59

M⁺ 74 (less than 10)

m/e

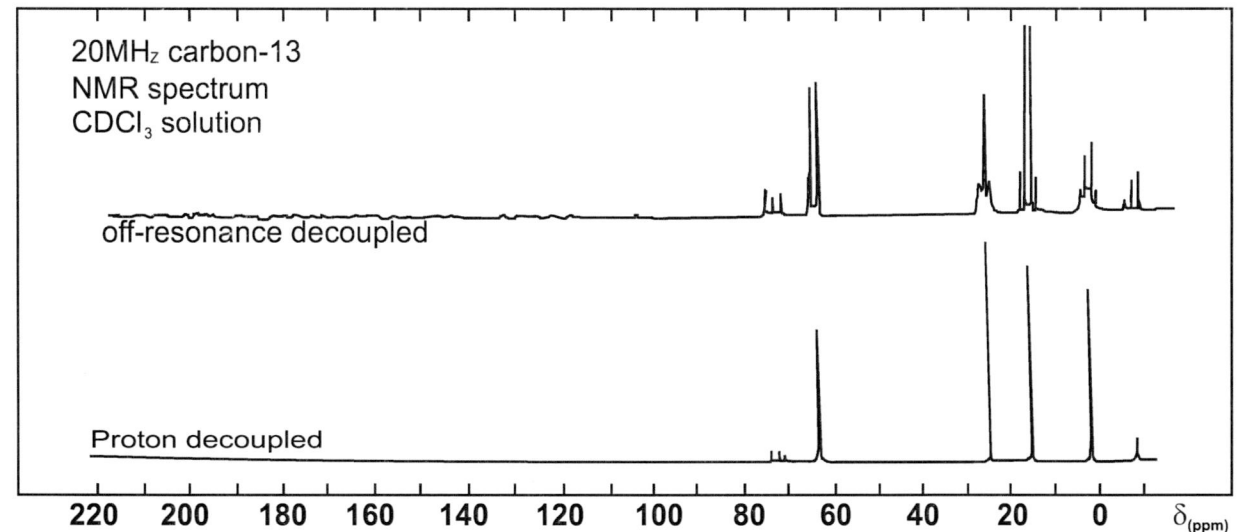

20MHz carbon-13
NMR spectrum
CDCl₃ solution

off-resonance decoupled

Proton decoupled

δ(ppm)

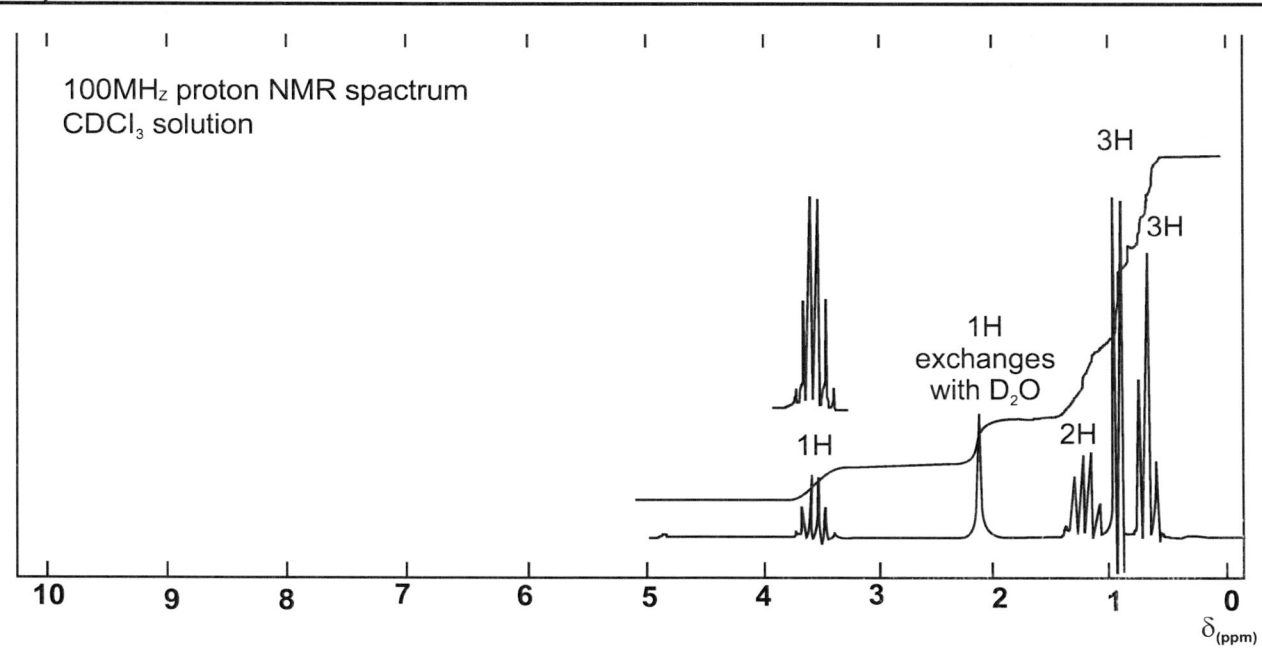

100MH_z proton NMR spactrum
CDCl₃ solution

3H

3H

1H
exchanges
with D₂O

1H

2H

δ_(ppm)

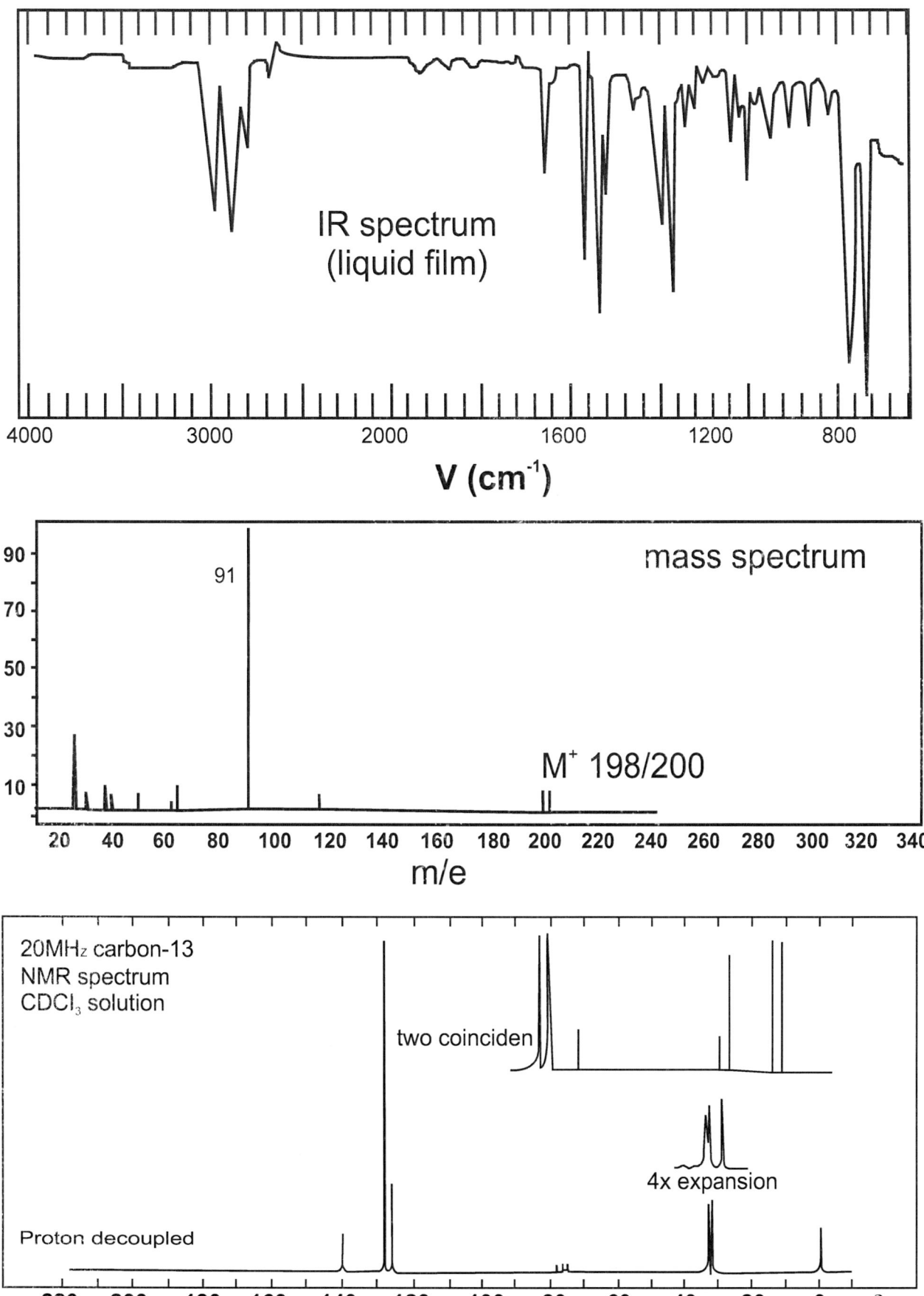

IR spectrum
(liquid film)

$V (cm^{-1})$

mass spectrum

91

M^+ 198/200

m/e

20MHz carbon-13
NMR spectrum
$CDCl_3$ solution

two coinciden

4x expansion

Proton decoupled

$\delta_{(ppm)}$

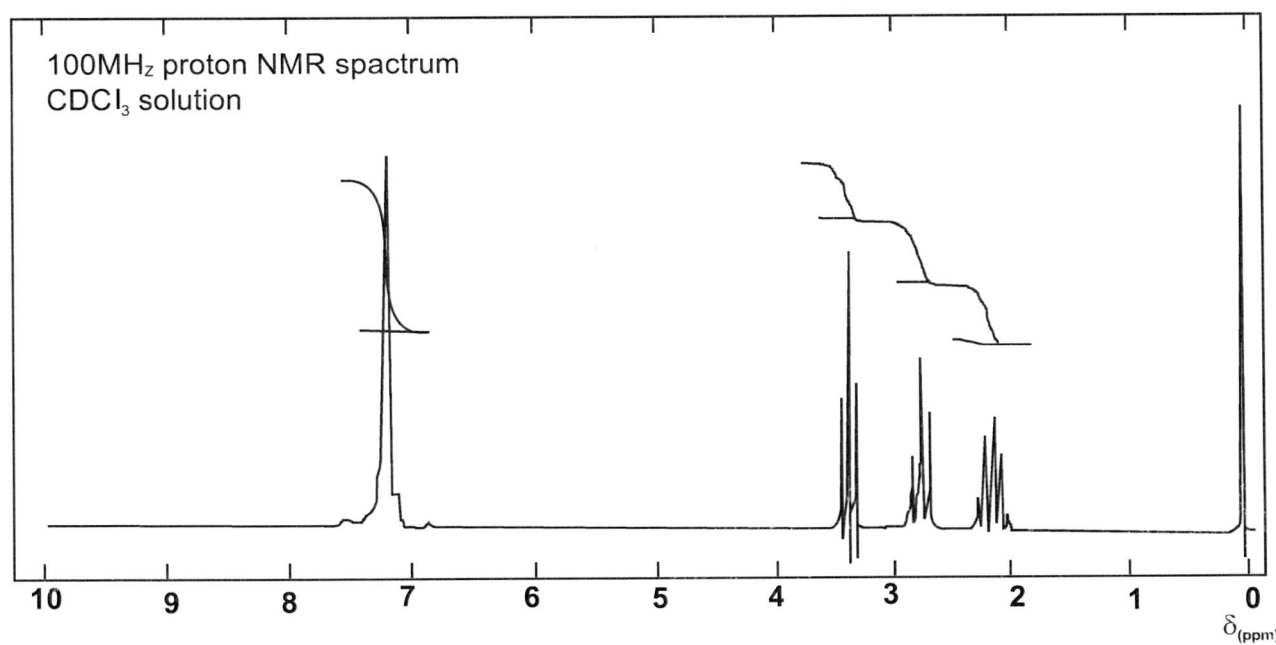

100MHz proton NMR spactrum
CDCl₃ solution

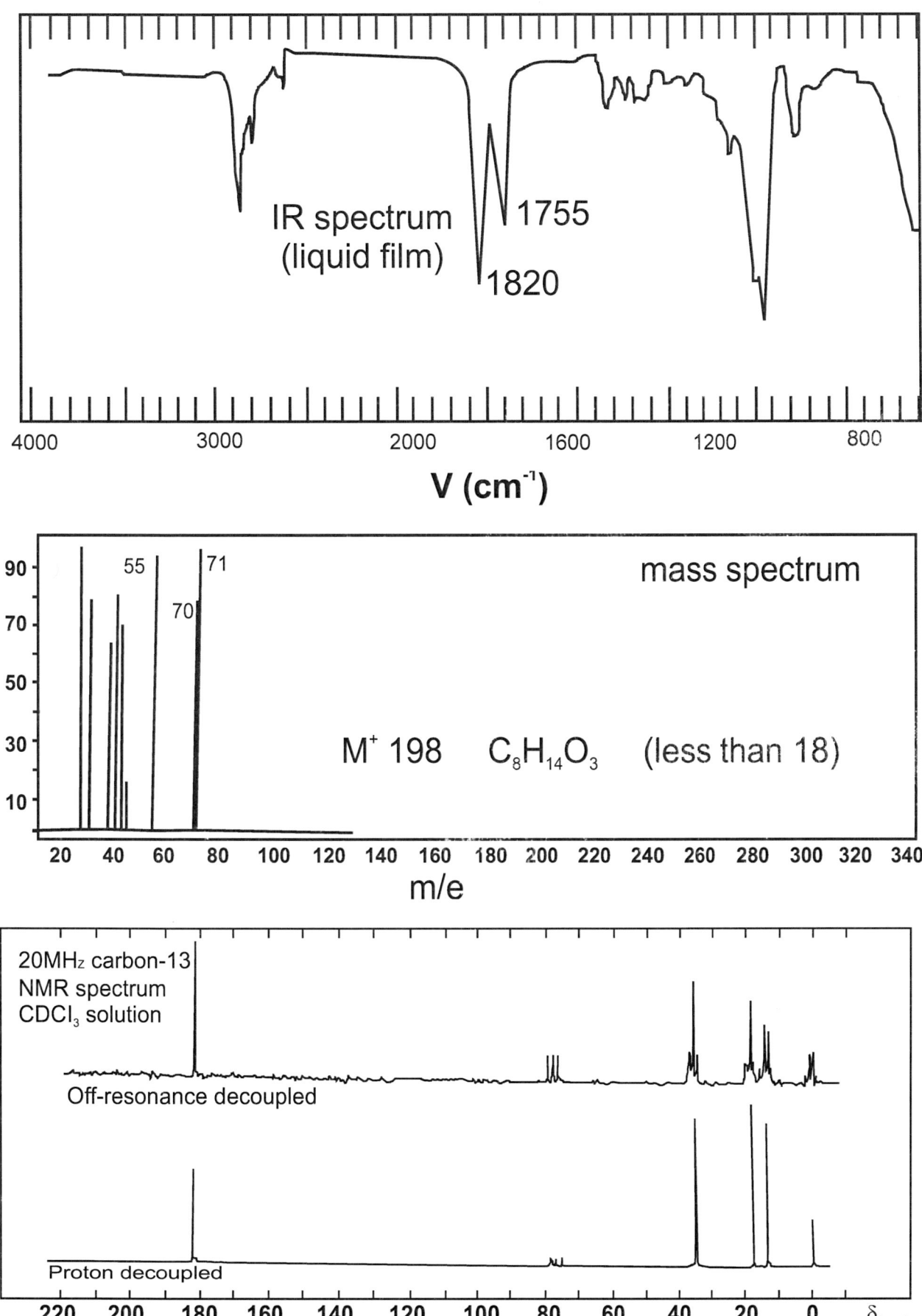

IR spectrum
(liquid film)

1755

1820

mass spectrum

90

70

50

30

10

55   70 71

M+ 198   C₈H₁₄O₃   (less than 18)

20 40 60 80 100 120 140 160 180 200 220 240 260 280 300 320 340
m/e

20MHz carbon-13
NMR spectrum
CDCl₃ solution

Off-resonance decoupled

Proton decoupled

220 200 180 160 140 120 100 80 60 40 20 0   δ(ppm)

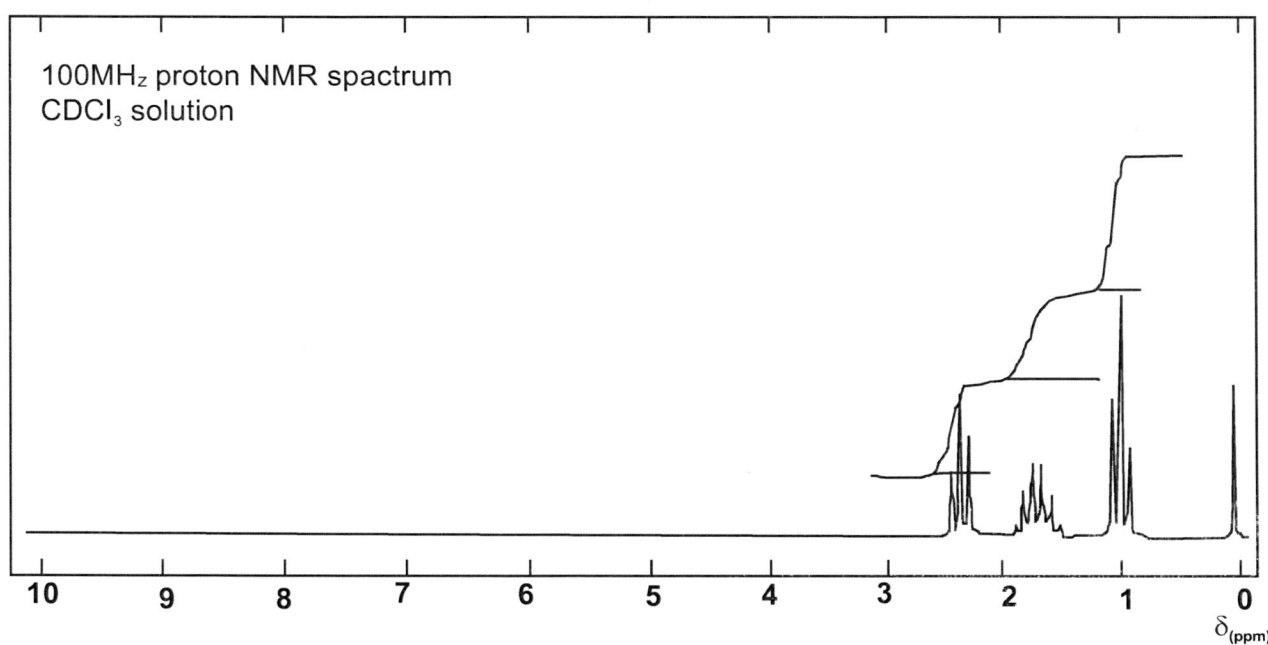

100MHz proton NMR spactrum
CDCl₃ solution

δ(ppm)

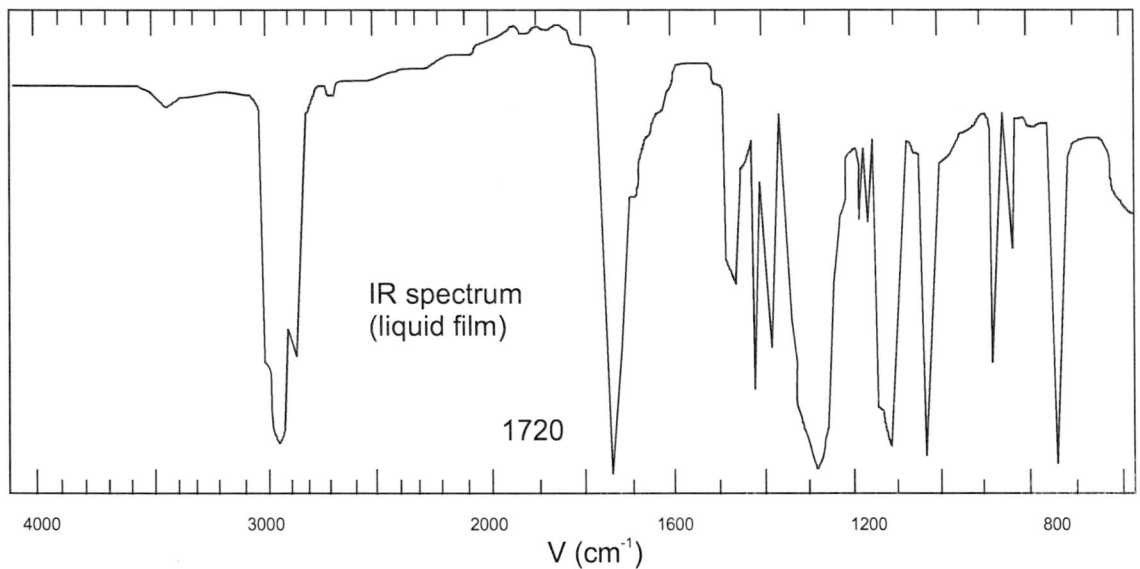

IR spectrum
(liquid film)

1720

V (cm⁻¹)

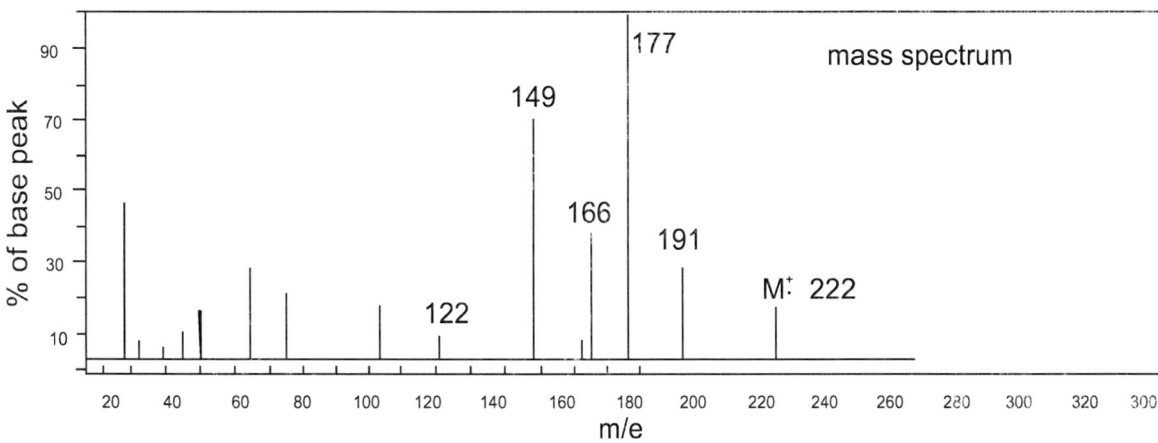

mass spectrum

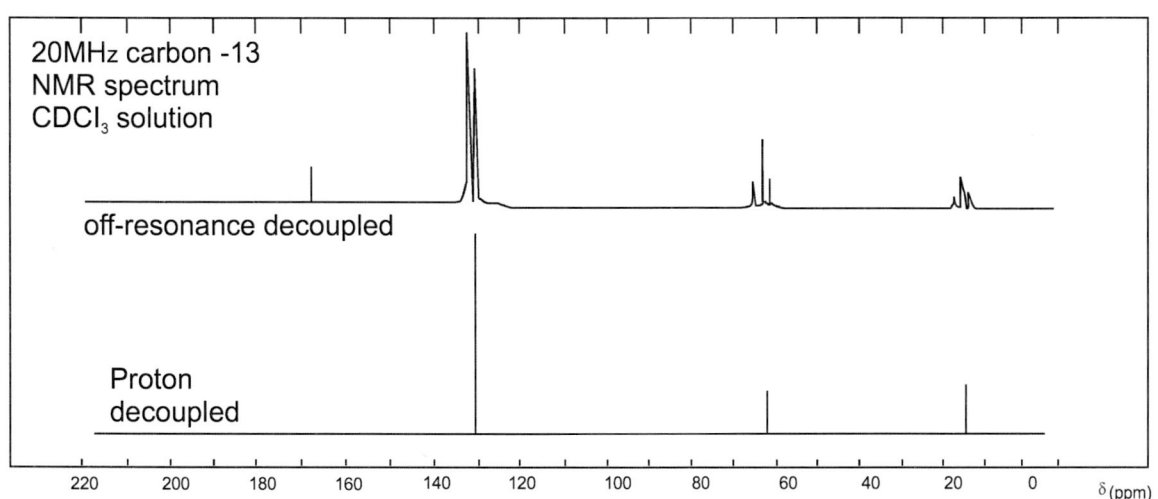

20MHz carbon -13
NMR spectrum
CDCl₃ solution

off-resonance decoupled

Proton
decoupled

δ (ppm)

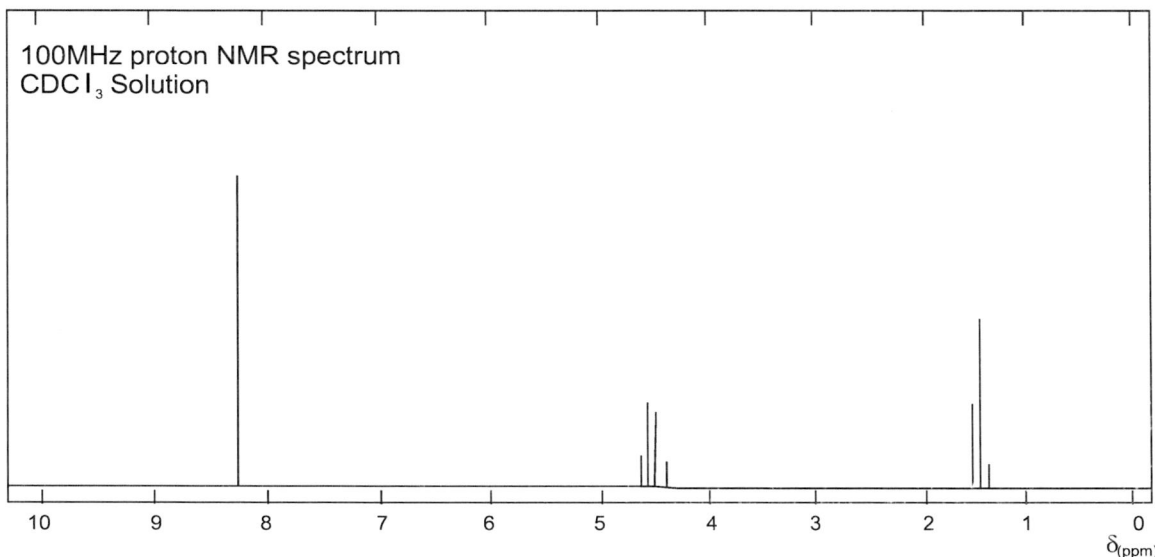

100MHz proton NMR spectrum
CDCl$_3$ Solution

$\delta_{(ppm)}$

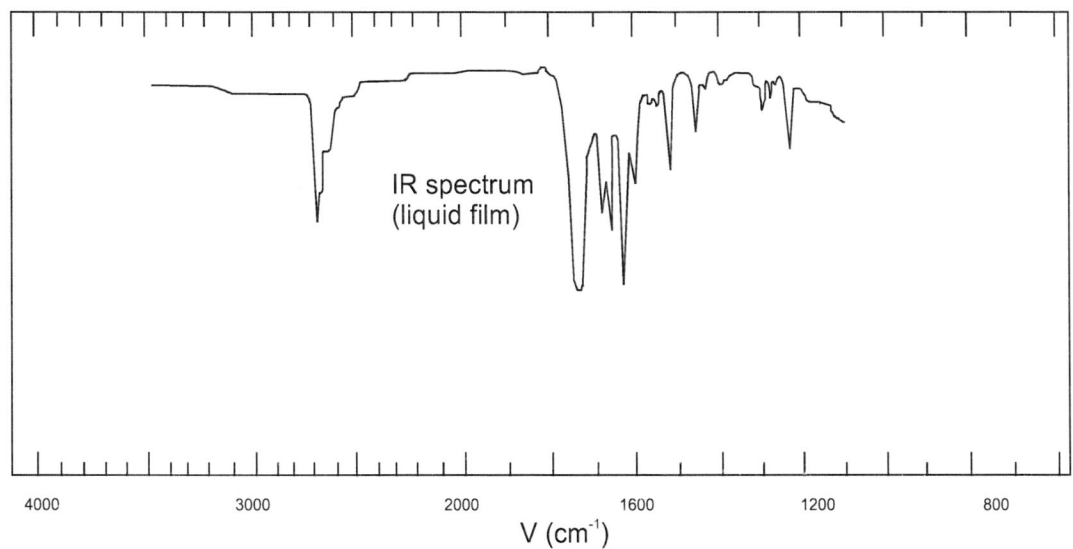

IR spectrum
(liquid film)

V (cm⁻¹)

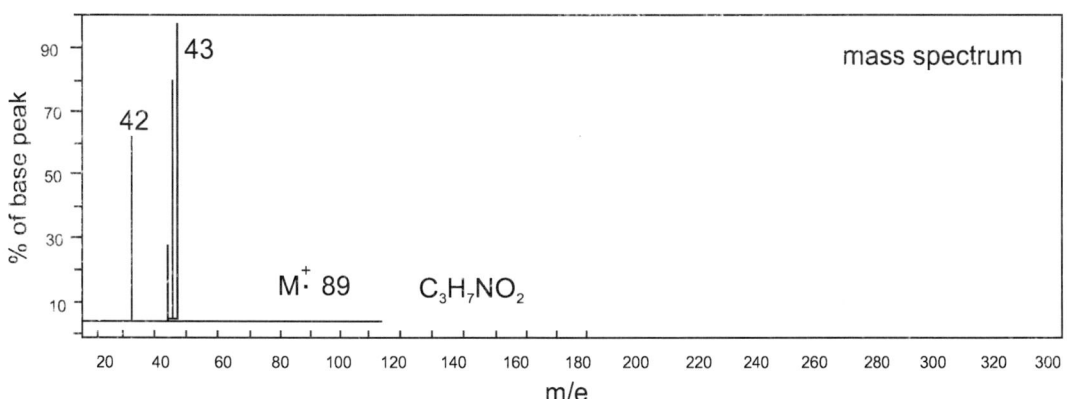

mass spectrum

43

42

M⁺ 89   C₃H₇NO₂

m/e

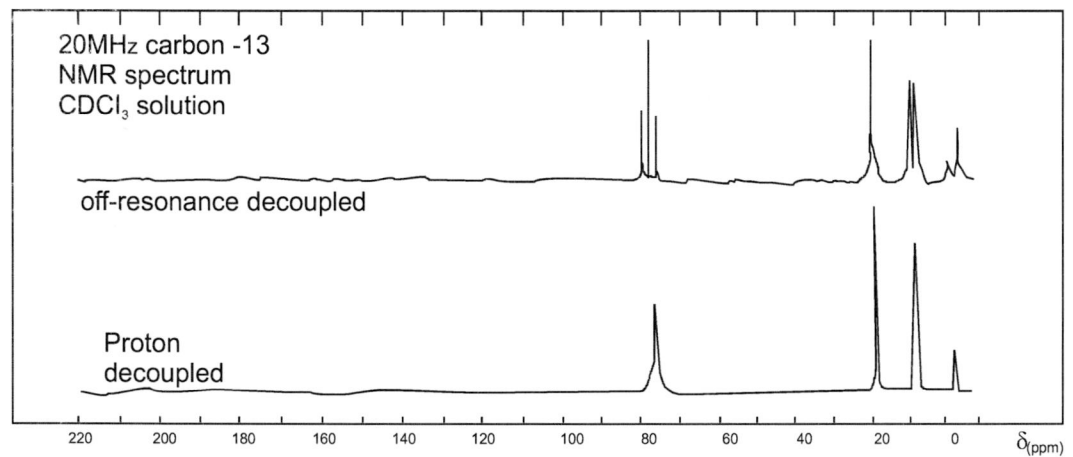

20MHz carbon -13
NMR spectrum
CDCl₃ solution

off-resonance decoupled

Proton
decoupled

δ(ppm)

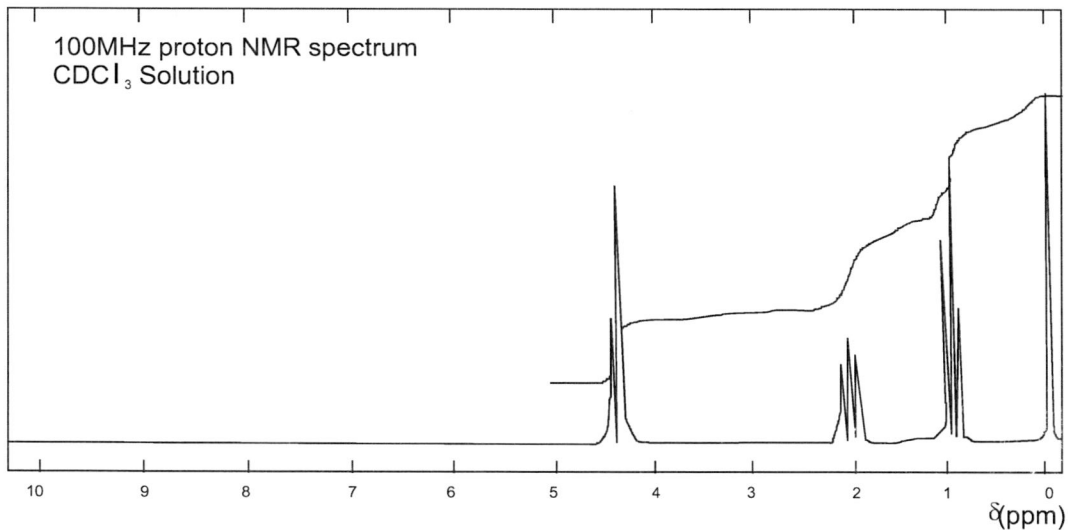

100MHz proton NMR spectrum
CDCl₃ Solution

δ(ppm)

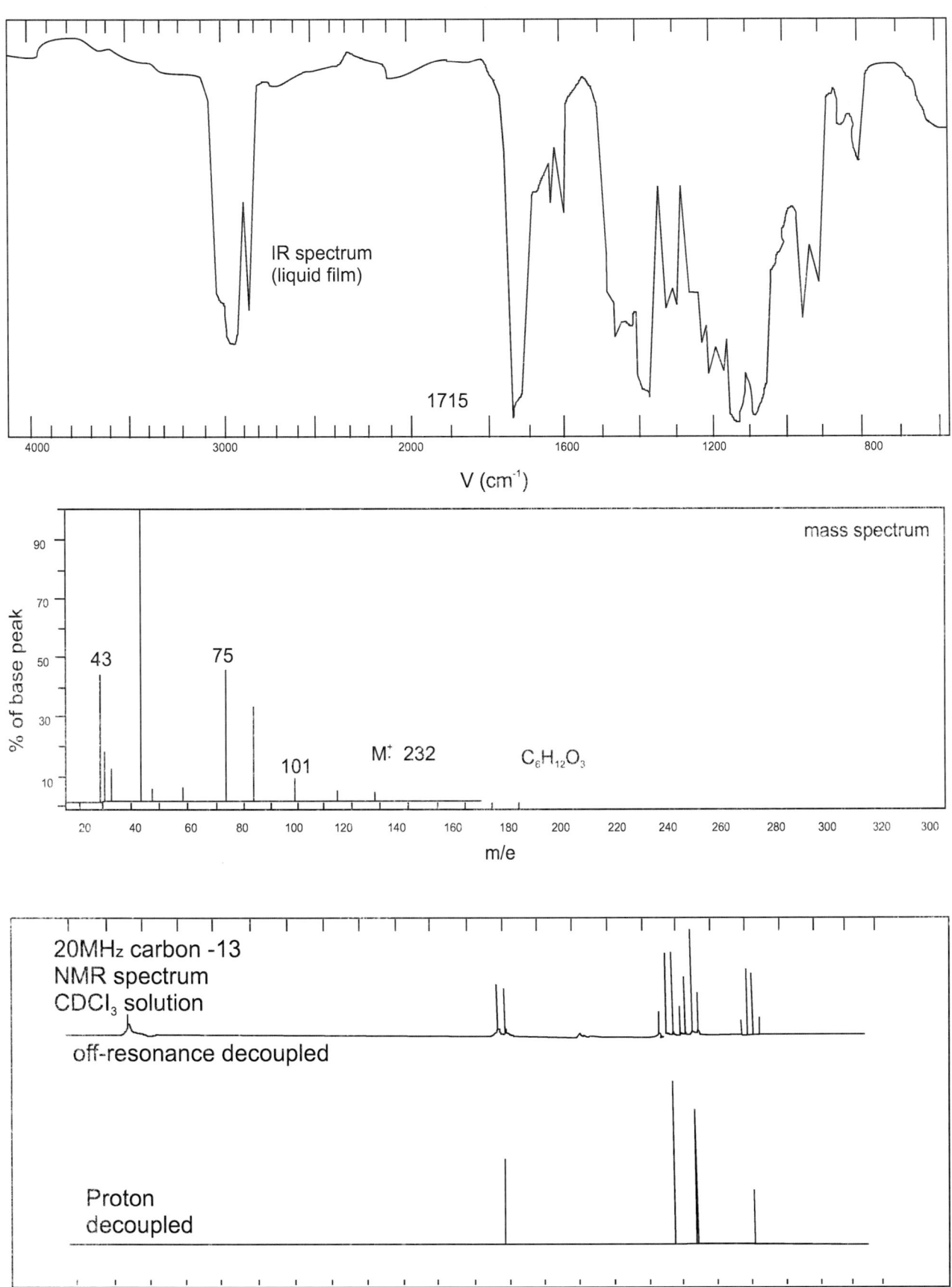

IR spectrum
(liquid film)

1715

mass spectrum

% of base peak

43

75

101

M⁺ 232

$C_6H_{12}O_3$

m/e

20MHz carbon -13
NMR spectrum
$CDCl_3$ solution

off-resonance decoupled

Proton
decoupled

$\delta_{(ppm)}$

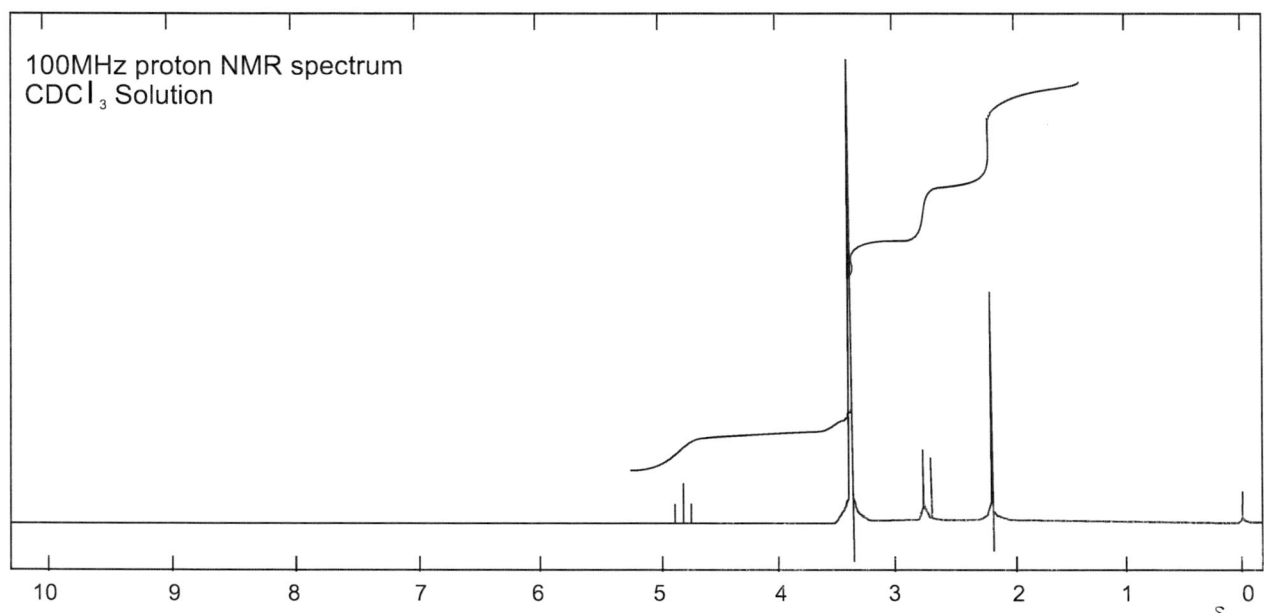

100MHz proton NMR spectrum
CDCl$_3$ Solution

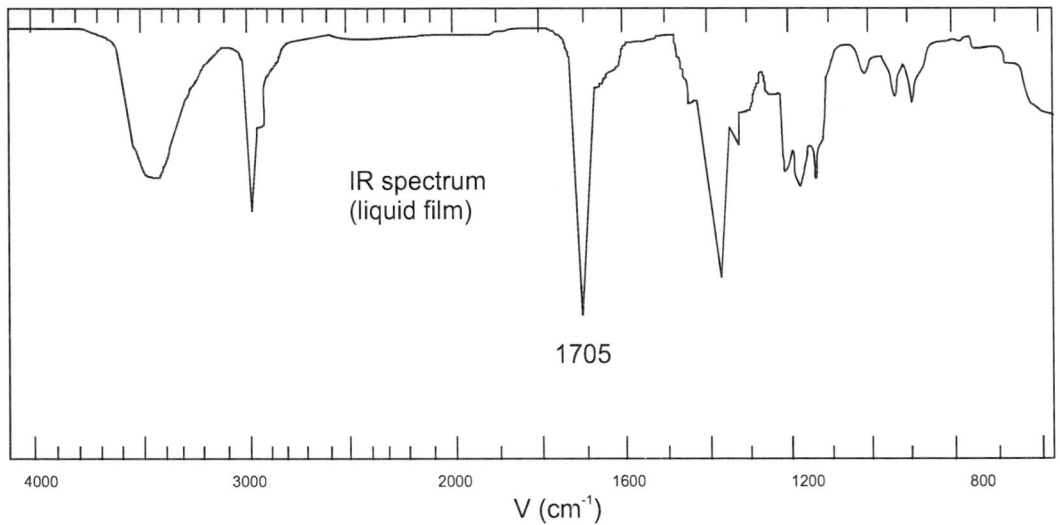

IR spectrum
(liquid film)

1705

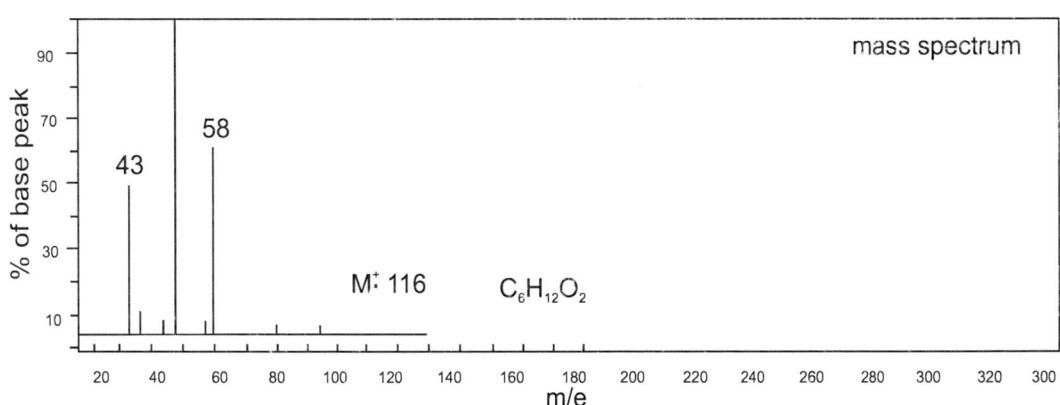

mass spectrum

% of base peak

90

70

50

43

30

58

10

M⁺ 116   C₆H₁₂O₂

20 40 60 80 100 120 140 160 180 200 220 240 260 280 300 300

m/e

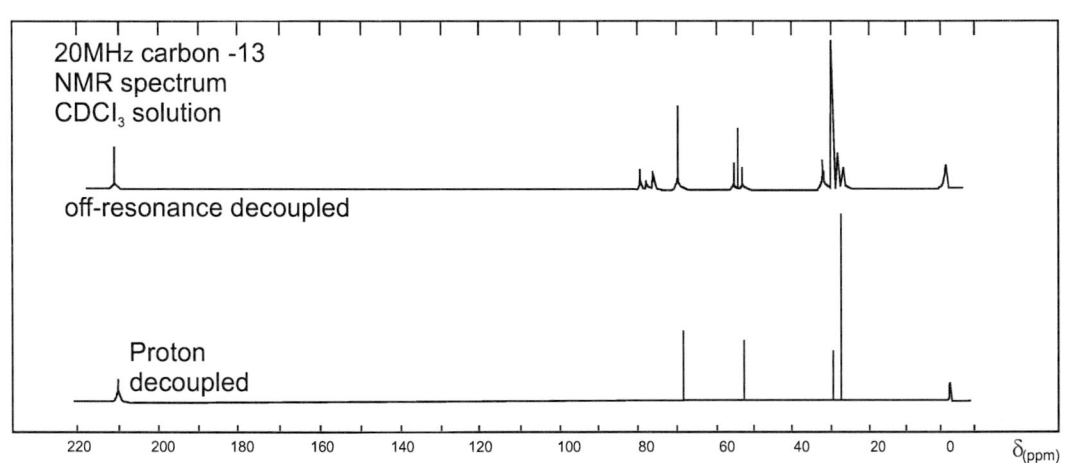

20MHz carbon -13
NMR spectrum
CDCl₃ solution

off-resonance decoupled

Proton
decoupled

220 200 180 160 140 120 100 80 60 40 20 0   δ(ppm)

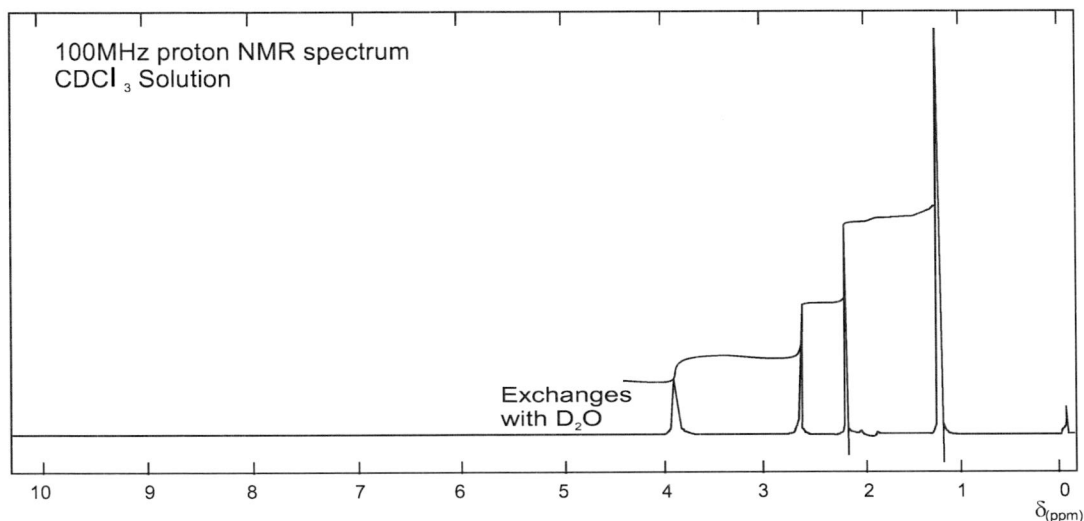

100MHz proton NMR spectrum
CDCl$_3$ Solution

Exchanges
with D$_2$O

$\delta_{(ppm)}$

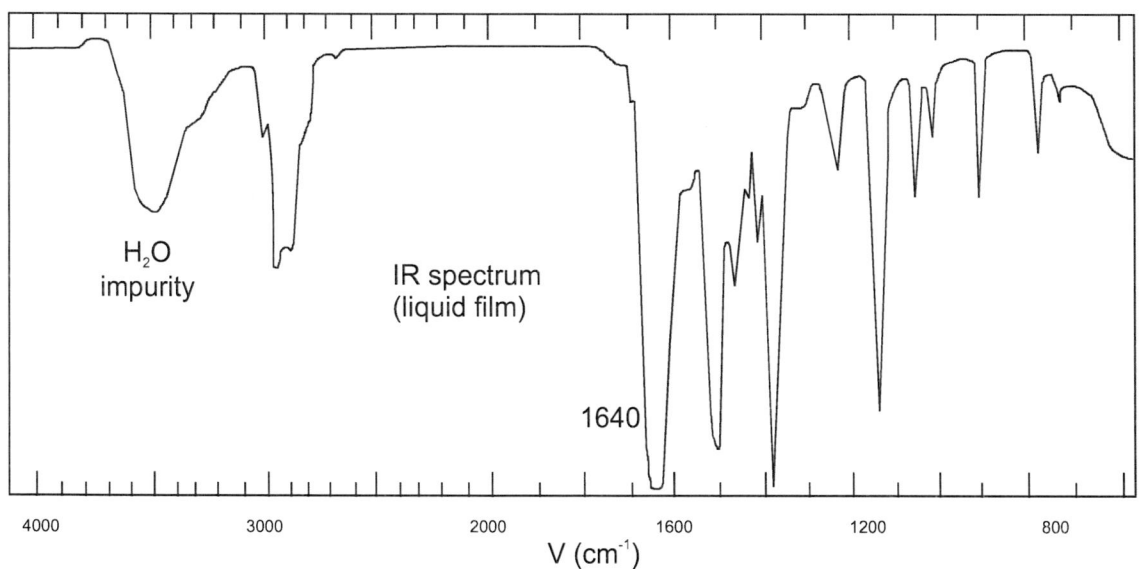

H₂O
impurity

IR spectrum
(liquid film)

1640

V (cm⁻¹)

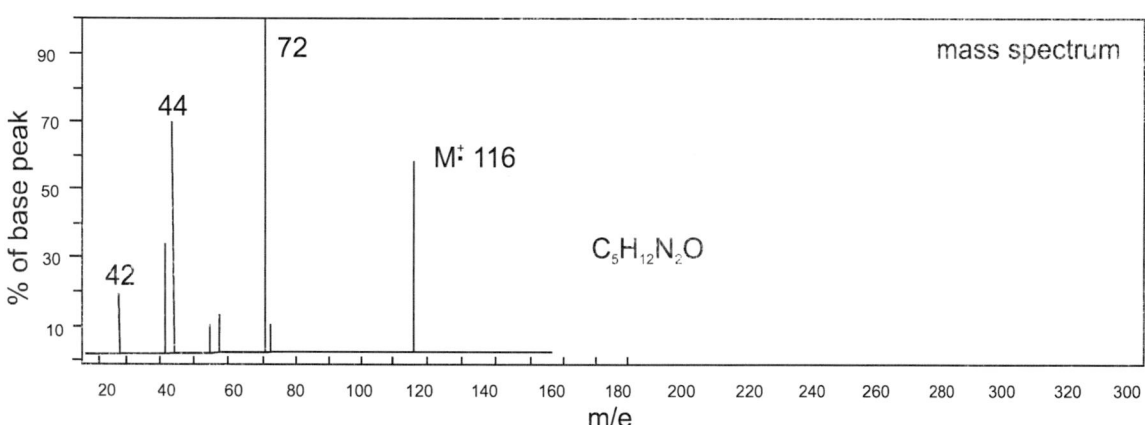

90

70

% of base peak

50

30

10

42

44

72

M⁺ 116

mass spectrum

$C_5H_{12}N_2O$

20   40   60   80   100   120   140   160   180   200   220   240   260   280   300   320   300

m/e

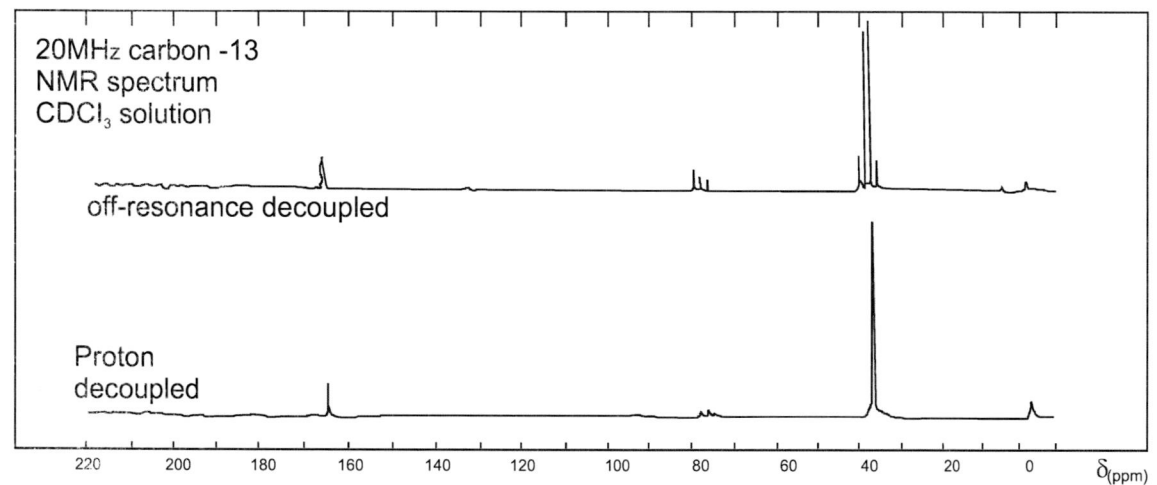

20MHz carbon -13
NMR spectrum
CDCl₃ solution

off-resonance decoupled

Proton
decoupled

220   200   180   160   140   120   100   80   60   40   20   0   δ(ppm)

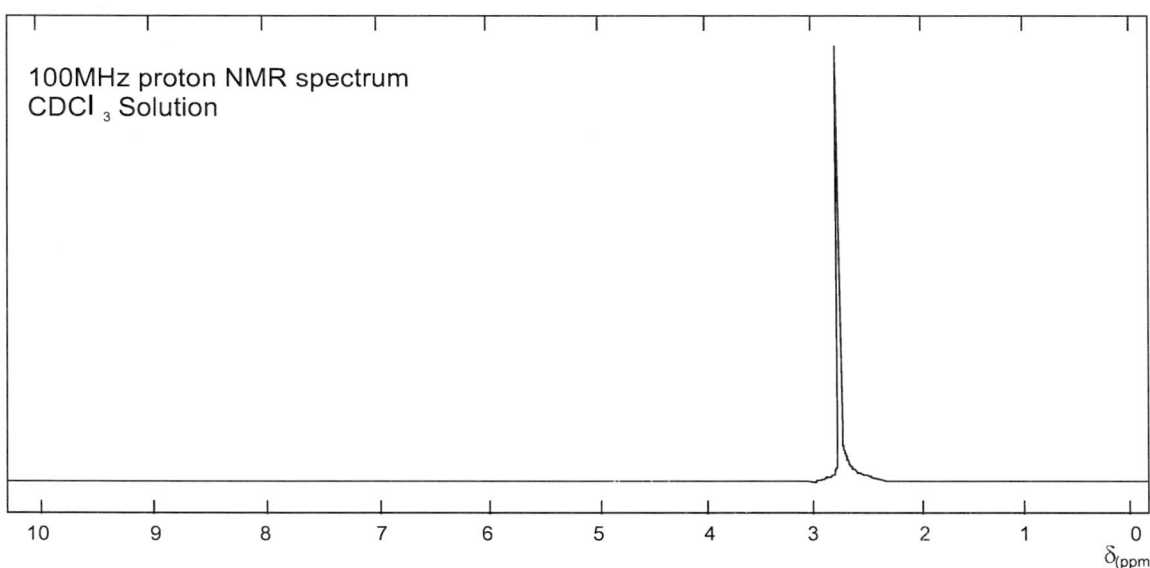

100MHz proton NMR spectrum
CDCl $_3$ Solution

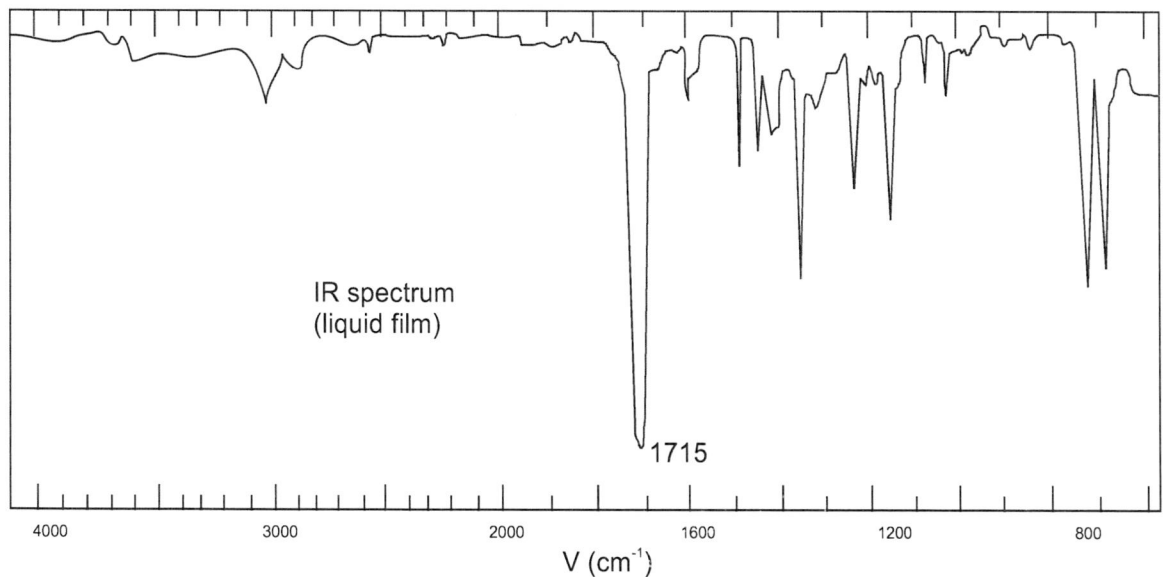

IR spectrum
(liquid film)

1715

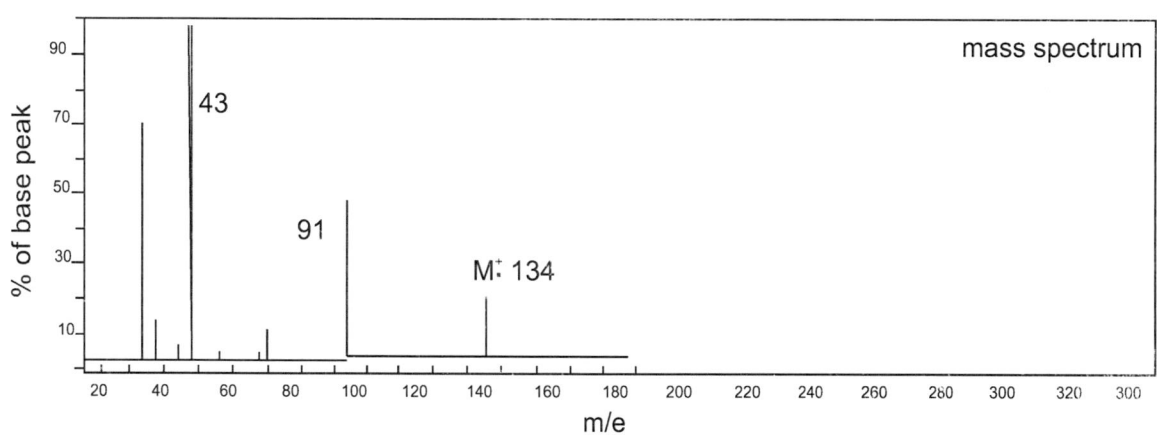

mass spectrum

% of base peak

43

91

M⁺ 134

m/e

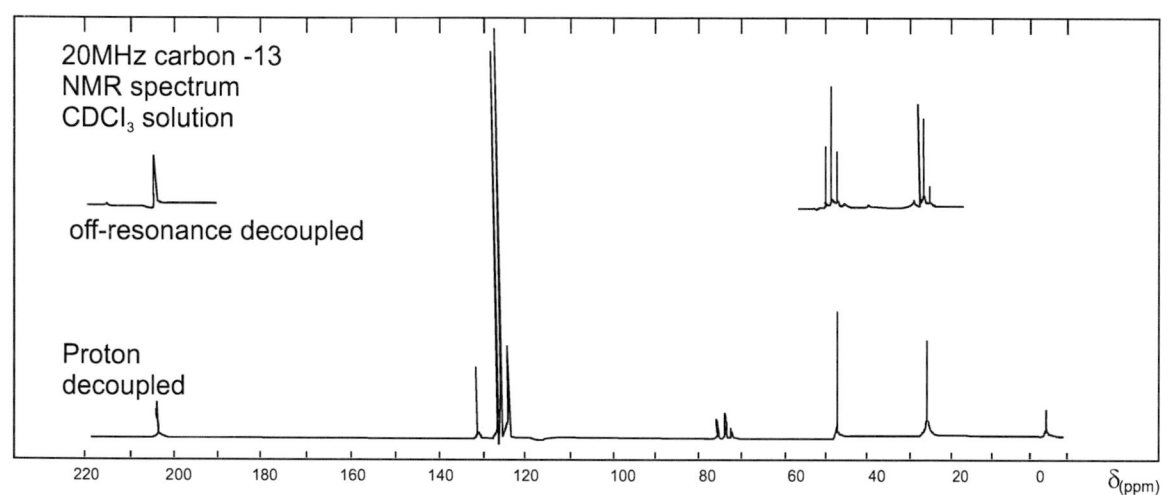

20MHz carbon -13
NMR spectrum
CDCl₃ solution

off-resonance decoupled

Proton
decoupled

δ(ppm)

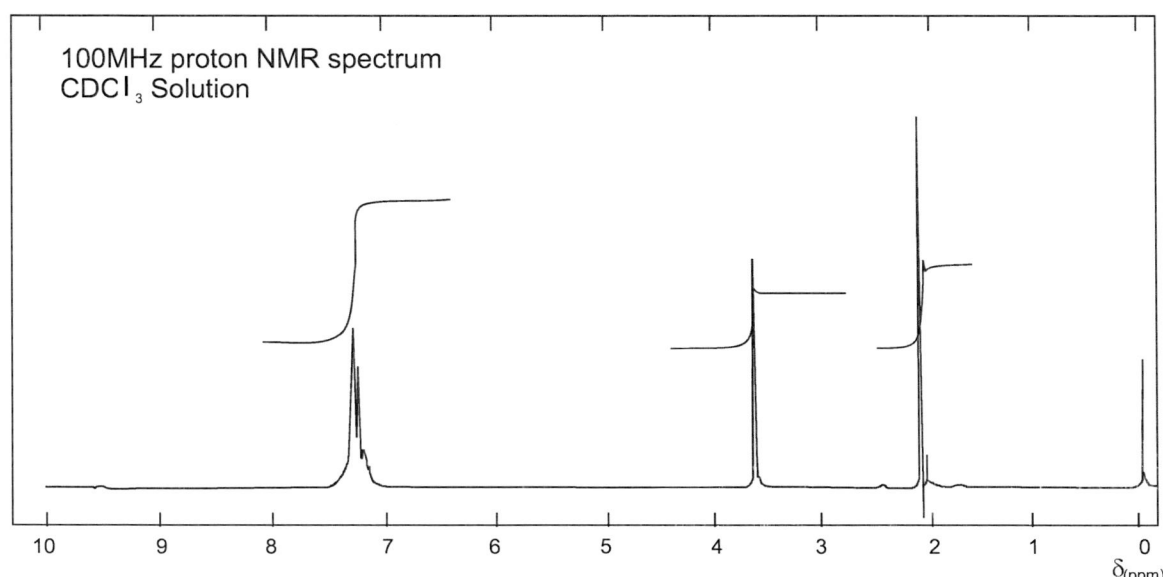

100MHz proton NMR spectrum
CDCl₃ Solution

δ(ppm)

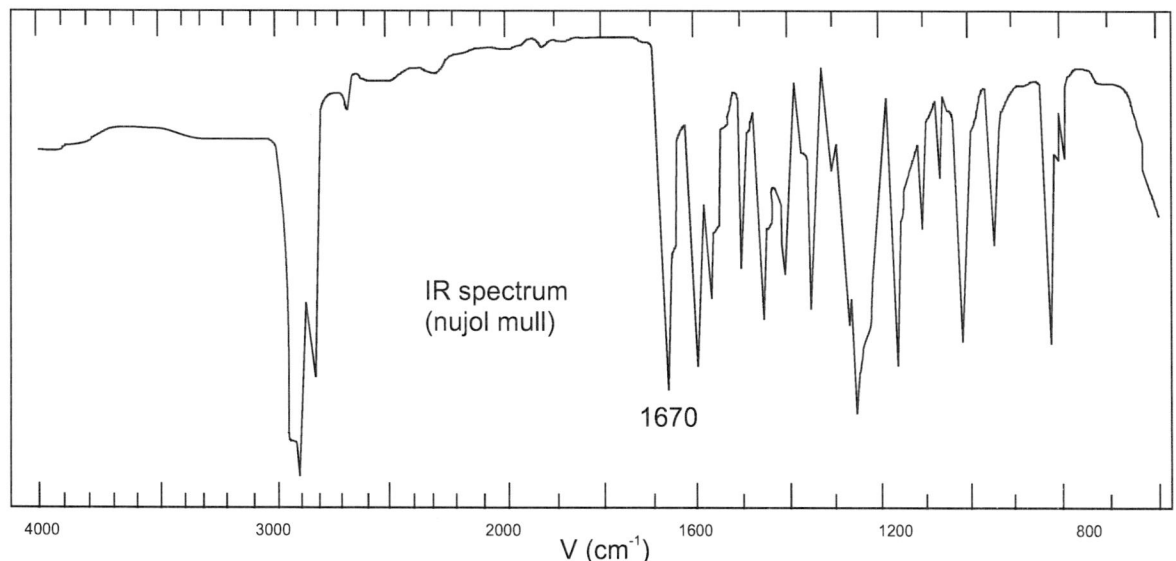

IR spectrum
(nujol mull)

1670

V (cm^{-1})

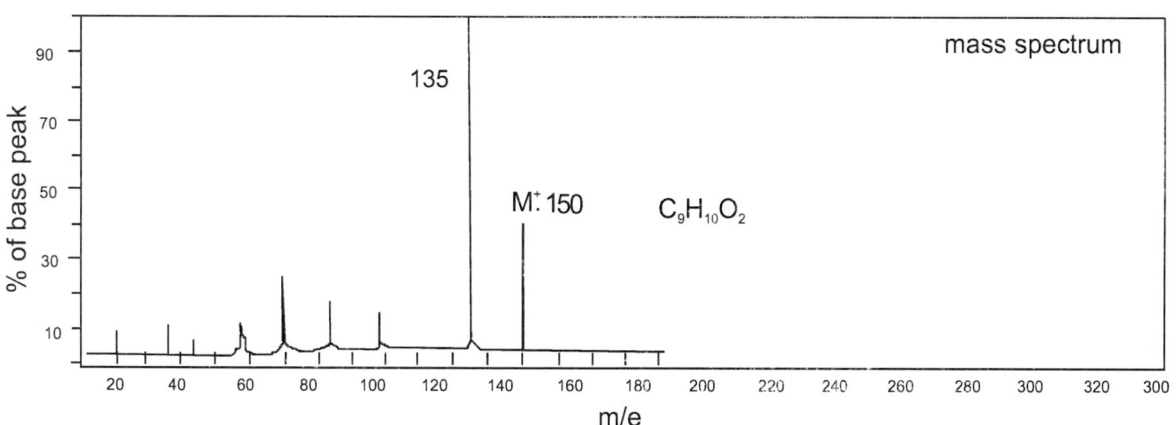

mass spectrum

135

M$^+$ 150

$C_9H_{10}O_2$

% of base peak

m/e

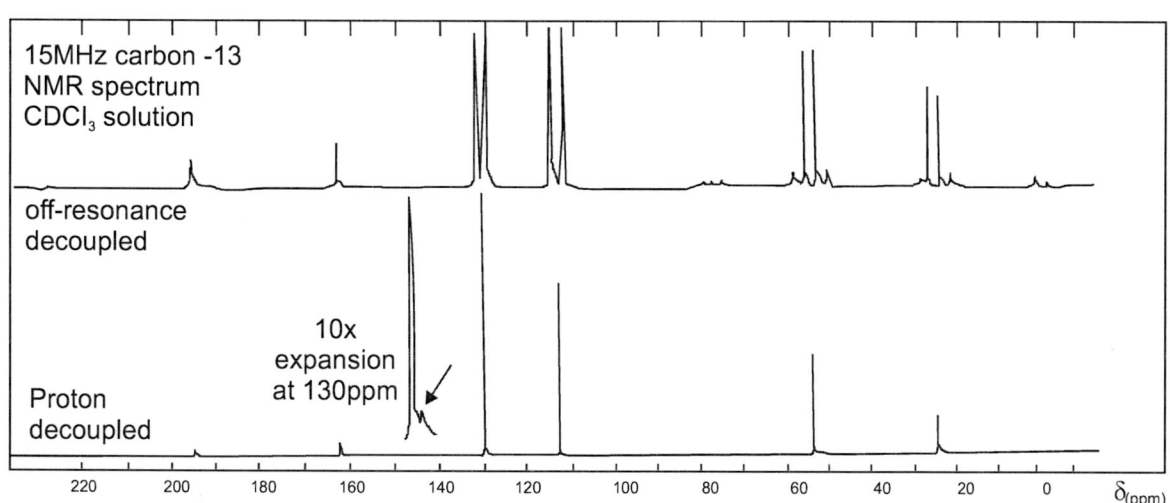

15MHz carbon -13
NMR spectrum
CDCl$_3$ solution

off-resonance
decoupled

10x
expansion
at 130ppm

Proton
decoupled

δ(ppm)

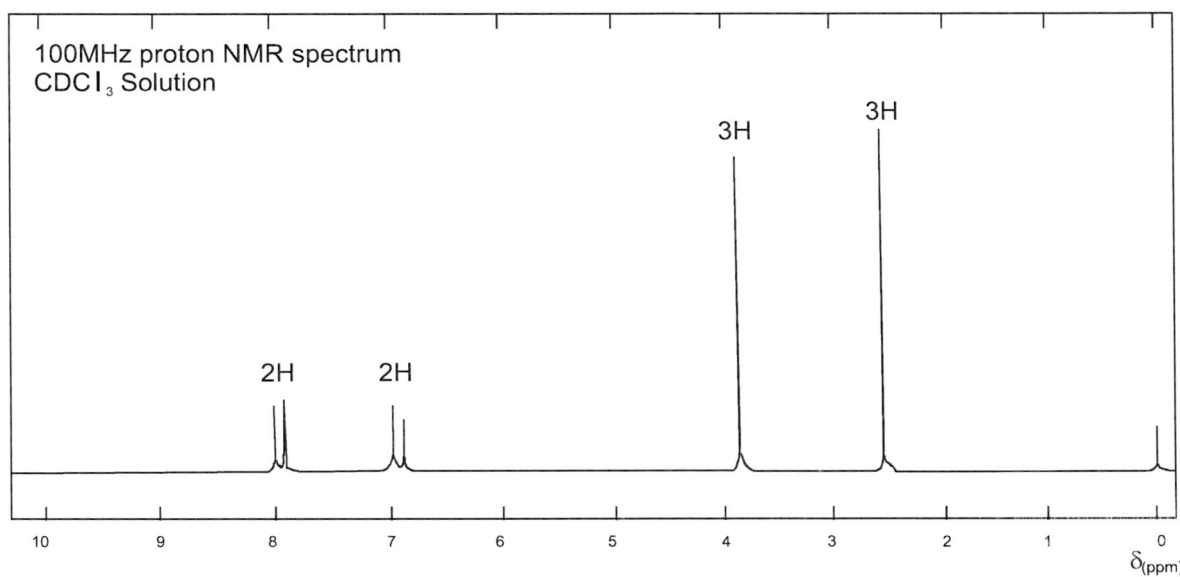

100MHz proton NMR spectrum
CDCl$_3$ Solution

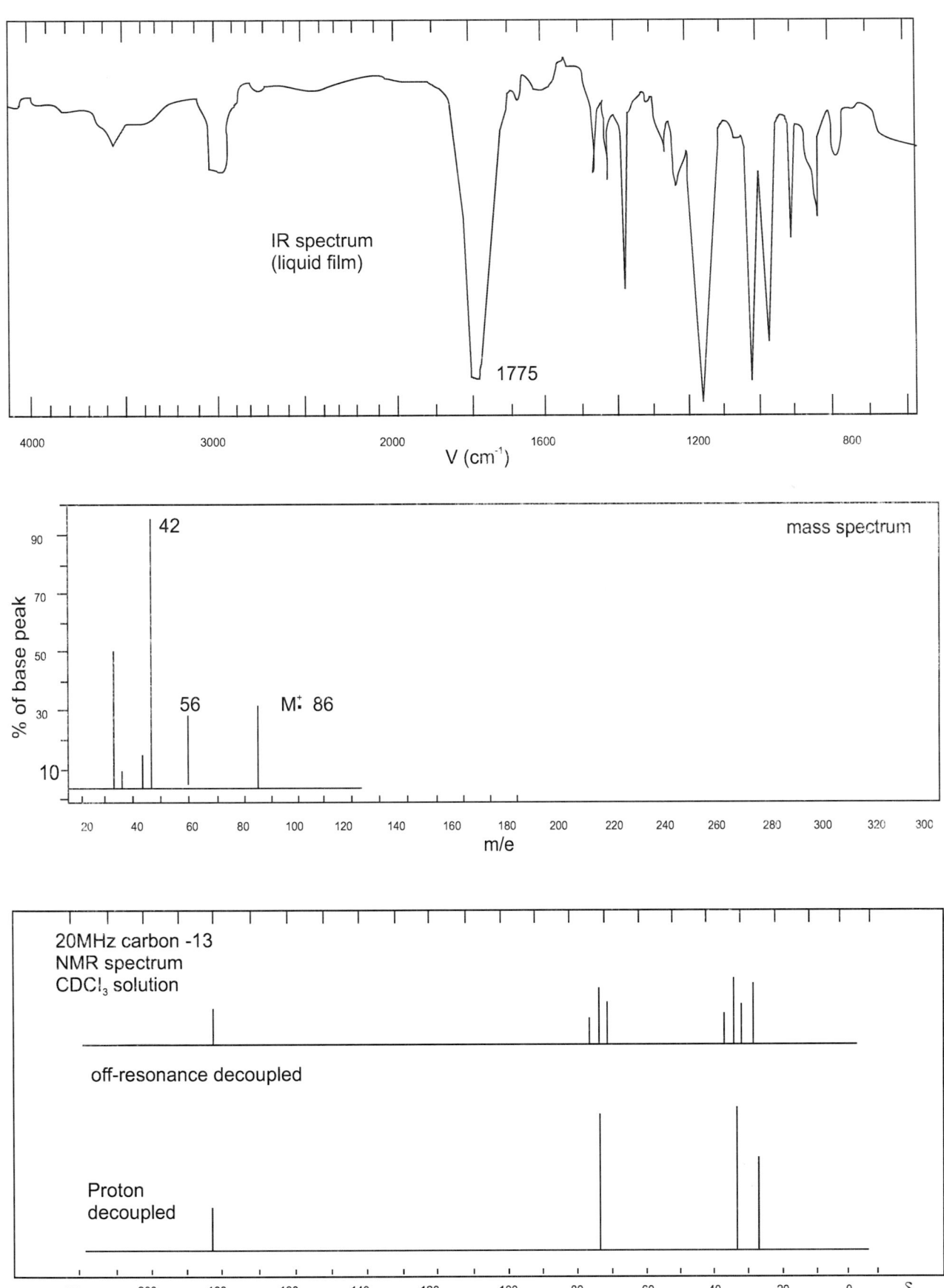

IR spectrum
(liquid film)

1775

mass spectrum

% of base peak

42

56

M⁺ 86

m/e

20MHz carbon -13
NMR spectrum
CDCl₃ solution

off-resonance decoupled

Proton
decoupled

δ(ppm)

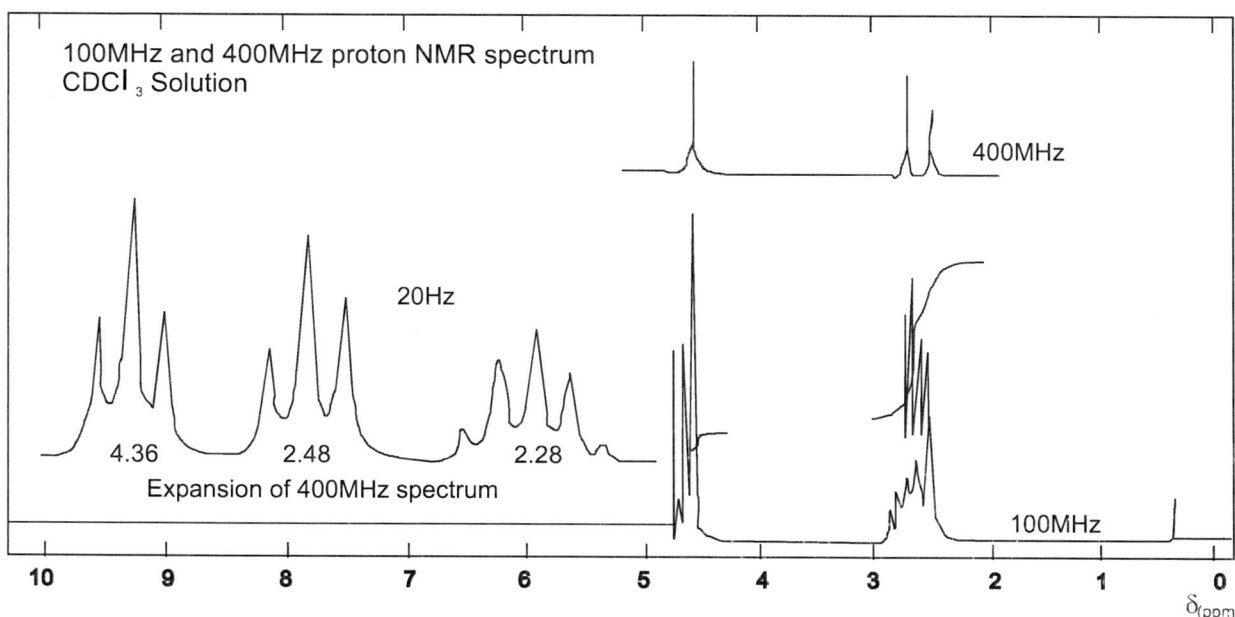

100MHz and 400MHz proton NMR spectrum
CDCl $_3$ Solution

400MHz

20Hz

4.36   2.48   2.28

Expansion of 400MHz spectrum

100MHz

10   9   8   7   6   5   4   3   2   1   0

$\delta_{(ppm)}$

Ultraviolet Spectroscopy

The ordinary type of visible-ultraviolet spectrometer measures absorption in the 200-800 nm range. The visible range goes down to about 400 nm, and from 400 to 200 nm is referred to as the ultraviolet. Whereas an infrared spectrum shows many sharp bands, an ultraviolet spectrum shows few bands and they are usually broad. This broadness is due to the fact that in a transition to a higher electronic level a molecule can go from any number of energy sub-levels (corresponding to various vibrational and rotational states) to any of a number of higher energy levels.

The customary way of describing a peak in the UV or visible spectrum is in terms of the position of the top of the band (λ max) and the intensity of that absorption (ε max or extinction coefficient). The magnitude of the ε max is directly proportional to the probability of the particular electronic transition which has caused the energy absorption.

The electronic transitions ($\rightarrow$) that occur in the ultraviolet and visible regions are of the following types:

$\sigma \rightarrow \sigma^*$

$n \rightarrow \sigma^*$

$n \rightarrow \pi^*$

$\pi \rightarrow \pi^*$

The $\sigma \rightarrow \sigma^*$ transition requires such a great amount of energy that it does not take place in the ordinary ultraviolet. As a consequence, compounds in which all valence shell electrons are involved in single bond formation do not show absorption except in the vacuum ultraviolet (below 200 nm).

Compounds that contain non-bonding electrons on oxygen, nitrogen, sulfur, or halogen may undergo n $\rightarrow \sigma^*$ transitions. For example methanol shows λ max at 183 nm (ε 150) and trimethylamine has λ max at 227 nm (ε 900).

Transitions which are of the greatest concern to organic chemists are n $\rightarrow \pi^*$ and

$\pi \rightarrow \pi^*$. Then $\rightarrow \pi^*$ transition involves the excitation of an electron of an un shared pair to an unstable (anti bonding) π orbital.

The $\pi \rightarrow \pi^*$ transition involves the moving of an electron from a stable (bonding) π orbital to an unstable (anti bonding) π^* orbital.

The $\pi \rightarrow \pi^*$ transition can occur in simple alkenes but the absorption is in the far ultraviolet and outside the range of the usual spectrophotometers. The figure below shows the relative amounts of energy required for the various transitions discussed.

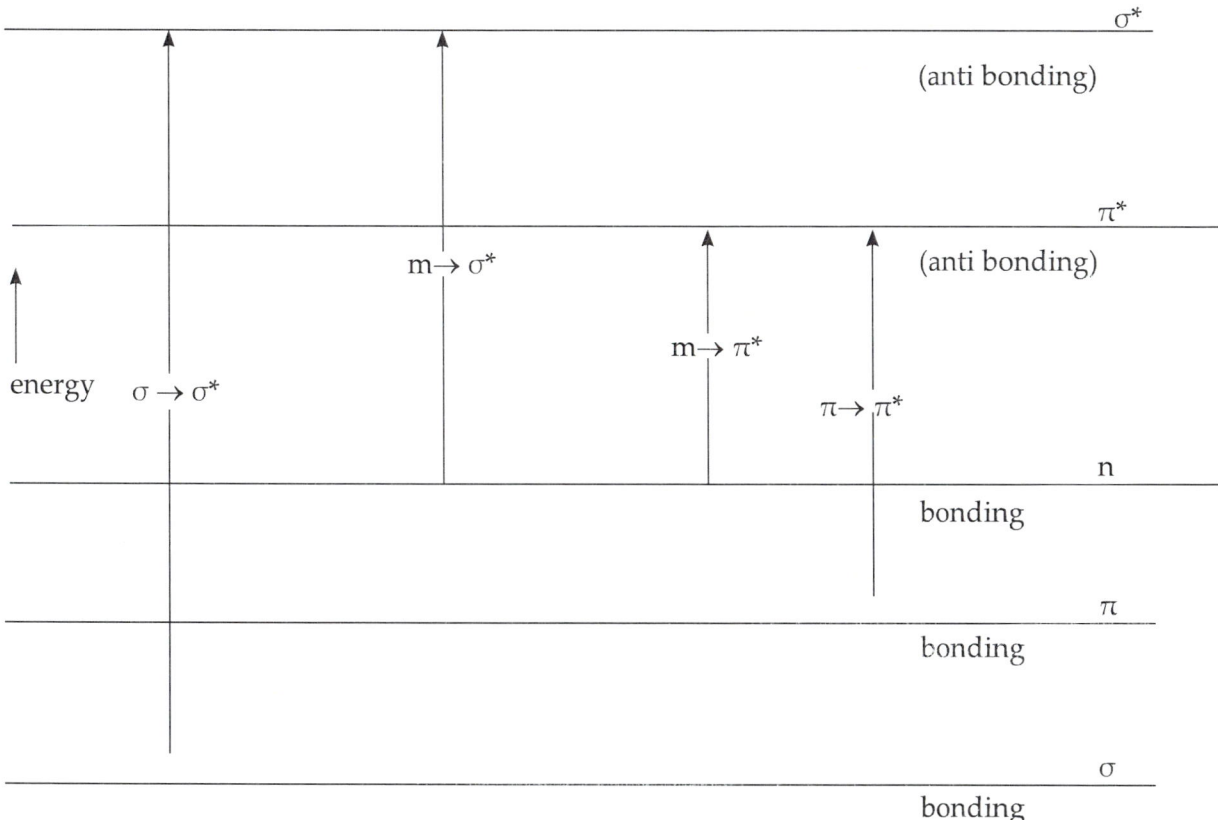

Energy transitions

Since even the n → π * and π→π *transitions at frequencies usually below 200 nm in isolated double bonds, ultraviolet is not too useful. Nevertheless, the situation is different when there are two or more conjugated double bonds, as in dienes, γ, B-unsaturated aldehydes and ketones, etc. When two chromophoric groups are conjugated, the

π→π * transition is generally shifted 15-50 nm to a longer wavelength compared to the non-conjugated systems. The n → π * transition band is also usually shifted about 30 nm to a longer wavelength.

The conjugated systems absorb at lower frequencies (higher wavelengths) than do the unconjugated systems because resonance stabilizes the excited states more than the ground state. Consequently, the energy differences between ground and excited states is less in the conjugated than in the unconjugated systems.

Ultraviolet (UV) Instrument

Most ultraviolet and visible recording spectrophotometers record wavelength versus absorbance. The absorbance or optical density is defined by

$$A = \log \frac{Io}{I}$$

Where **Io** is the intensity of incident light and **I** is the intensity of emitted light. The range of absorbance is commonly 0 - 4. Calculation of the intensity of an absorption band involves the use of Beer's and Lambert's Laws. **Lambert's Law** states that the intensity of transmitted light through a homogeneous medium decreases geometrically as the thickness of the layer increases arithmetically. **Beer's Law** states that each molecule of solute absorbs the same fraction of incident light regardless of concentration in a non-absorbing medium. Deviations from Beer's Law are not great over the low concentration range used in ultraviolet spectroscopy. The above laws may be stated by the formula

$$\in = \frac{A}{cl}$$

Where $\in$ is the molar extinction coefficient, **c** is the molar concentration, and **l** is the path length in centimeters. Results are customarily reported in terms of $\in$ or **log** $\in$.

UV/VIS Instructions

1.   Turn on Lamp(s); allow to warm up for at least 10 minutes.

2.   Double Click on the UV/VIS icon on desktop.

3.   Select Cancel when prompted to enter operator name and password.

4.   Rinse cuvette at least 3 times with desired blank (in some cases water), fill cuvette ¾ full, carefully remove any air bubbles and wipe outside clean with a kim wipe.

5.   Place blank in slot #1.

6.   Repeat step four with a second cuvette and sample.

7.   Place Sample in slot #2.

8.   Turn lever on the Multicell sampler clockwise to lock cuvettes into place.

9.   On the left side of the computer screen select Multicell under sampling.

10.   Make sure Cuvette #1 is selected in the lower left corner of the screen and click Blank.

11.   Select Cuvette #2, and click sample to obtain the spectrum.

12.   To print your spectrum, select the spectrum window, go to File and select 'Print selected window'.

13.   When finished, close the UV/VIS Program, and be sure to TURN THE LAMPS OFF.

14.   Clean and return cuvettes to their case.

Sample of UV Spectrum

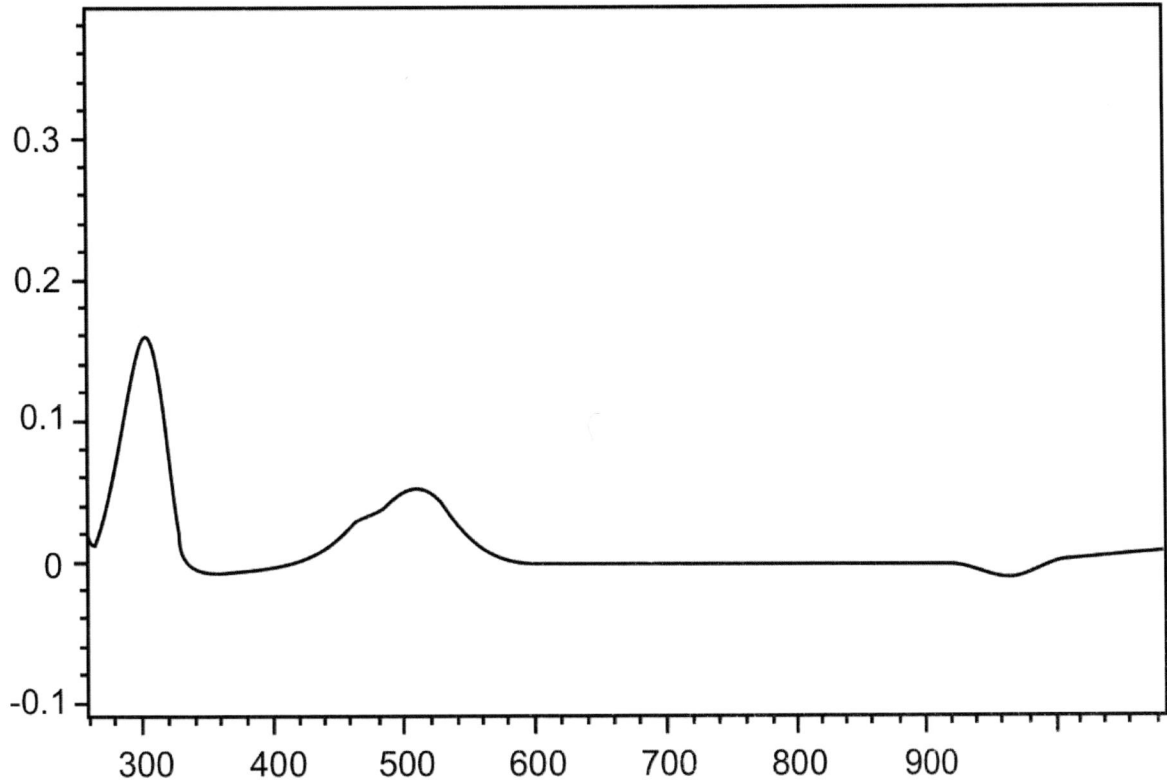

C hapter 10

MAKING A COMPOUND

In order to achieve the results of a particular experiment, and to ensure reproducibility, it is very important that the appropriate precautions are taken, and that proper preparations are made. Most organic reactions involve the use of anhydrous conditions, and as such, it is always advisable to use dry glassware for any type of reaction.

The most important thing before carrying out a reaction is that, the procedure must be read before going to the laboratory. Request for clarification during prelaboratory session. While in lab you have to make sure that you have the necessary glassware, reagents, solvents and apparatus. In addition to the above, always check on the safety aspects of the chemicals you are using for that particular experiment. Remember to write down your observations as the reaction progresses.

Synthesizing organic compounds follows a particular sequence: **The Reaction, Separation, Purification and Characterization**.

The Reaction: This is the most important part of performing an experiment because of the decisions on limiting reagent, method for adding reagents, drying and assembling glassware, reaction set-ups and protection of reagents and peers in the lab.

Separation (Isolation of the crude product): This is achieved by one of a variety of techniques (e.g. evaporation, filtration, concentration, extraction, distillation and precipitation). The experiment will determine the type of technique to be used.

Purification: This is the technique used to purify a product after it has been isolated from a reaction. Some of the purification techniques are: crystallization, distillation, chromatography, and sublimation.

Characterization: It is important to acquire as much information as possible on the product. The full set of routine physical data which could, and should be obtained on a pure compound are: m.p., b.p., ir, nmr(proton), uv, m.s. and gc.

FUME HOOD

Process Flow Steps in Drug Manufacturing Call for Different Analyses and Analytical Technologies.

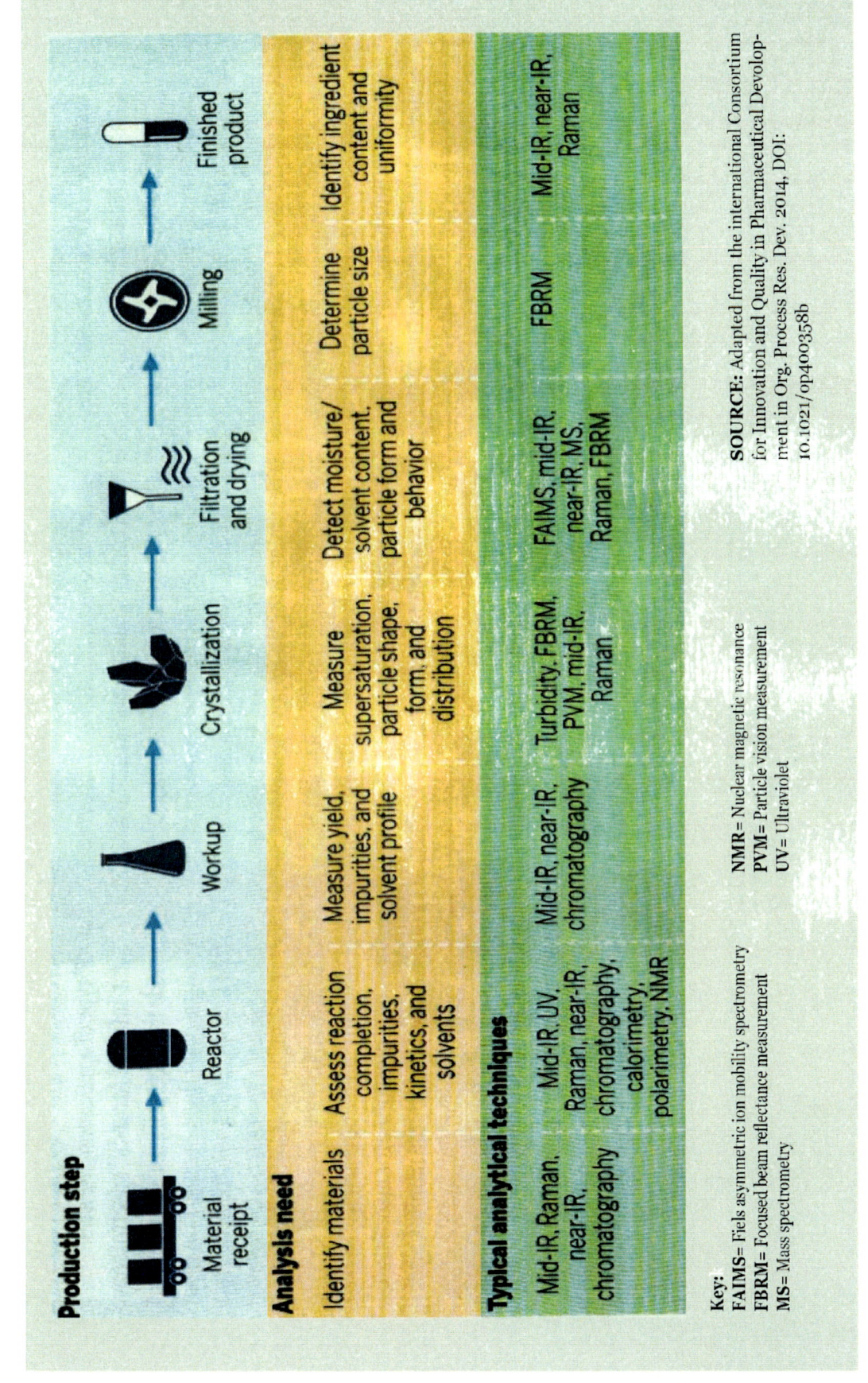

Production step

Material receipt → Reactor → Workup → Crystallization → Filtration and drying → Milling → Finished product

Analysis need

| Identify materials | Assess reaction completion, impurities, kinetics, and solvents | Measure yield, impurities, and solvent profile | Measure supersaturation, particle shape, form, and distribution | Detect moisture/solvent content, particle form and behavior | Determine particle size | Identify ingredient content and uniformity |

Typical analytical techniques

| Mid-IR, Raman, near-IR, chromatography | Mid-IR, UV, Raman, near-IR, chromatography, calorimetry, polarimetry, NMR | Mid-IR, near-IR, chromatography | Turbidity, FBRM, PVM, mid-IR, Raman | FAIMS, mid-IR, near-IR, MS, Raman, FBRM | FBRM | Mid-IR, near-IR, Raman |

Key:
FAIMS= Fiels asymmetric ion mobility spectrometry
FBRM= Focused beam reflectance measurement
MS= Mass spectrometry
NMR= Nuclear magnetic resonance
PVM= Particle vision measurement
UV= Ultraviolet

SOURCE: Adapted from the international Consortium for Innovation and Quality in Pharmaceutical Development in Org. Process Res. Dev. 2014, DOI: 10.1021/op400358b

Nitration Experiment

Figures show some glassware and set-up for some selected reactions

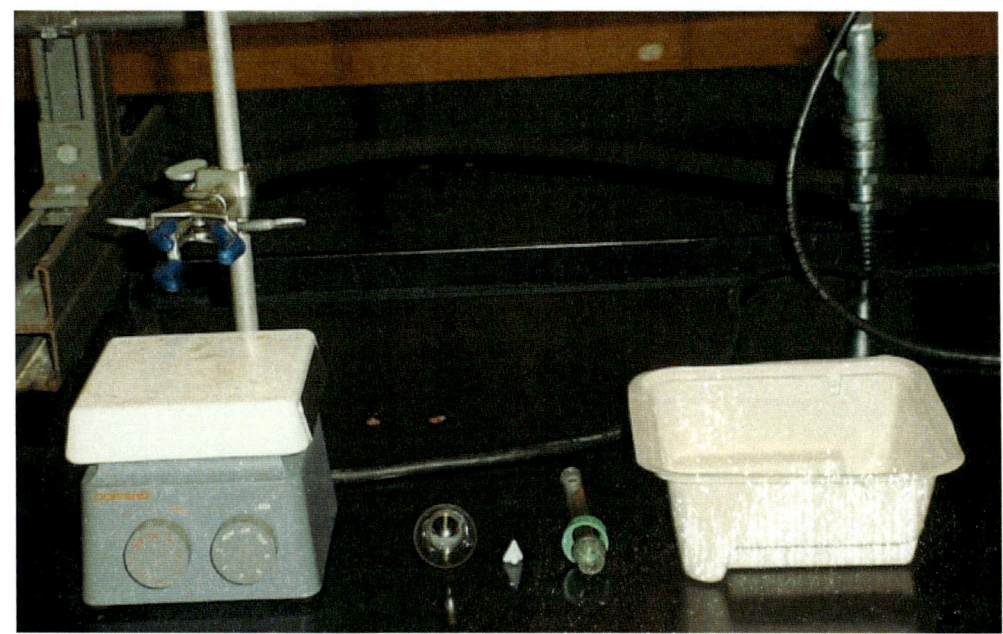

Equipments Needed

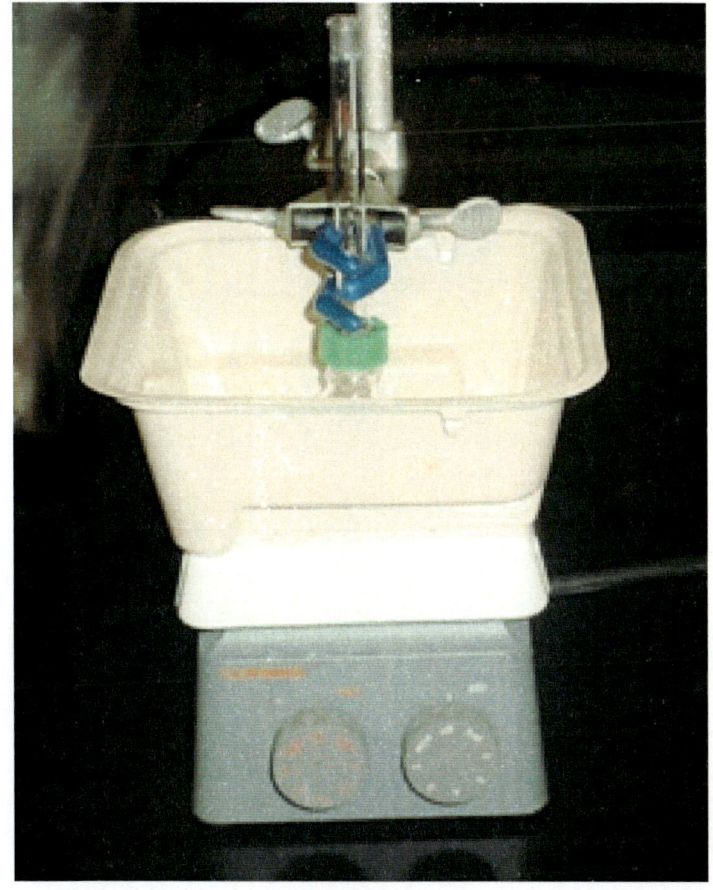

Set-up

Grignard Reaction

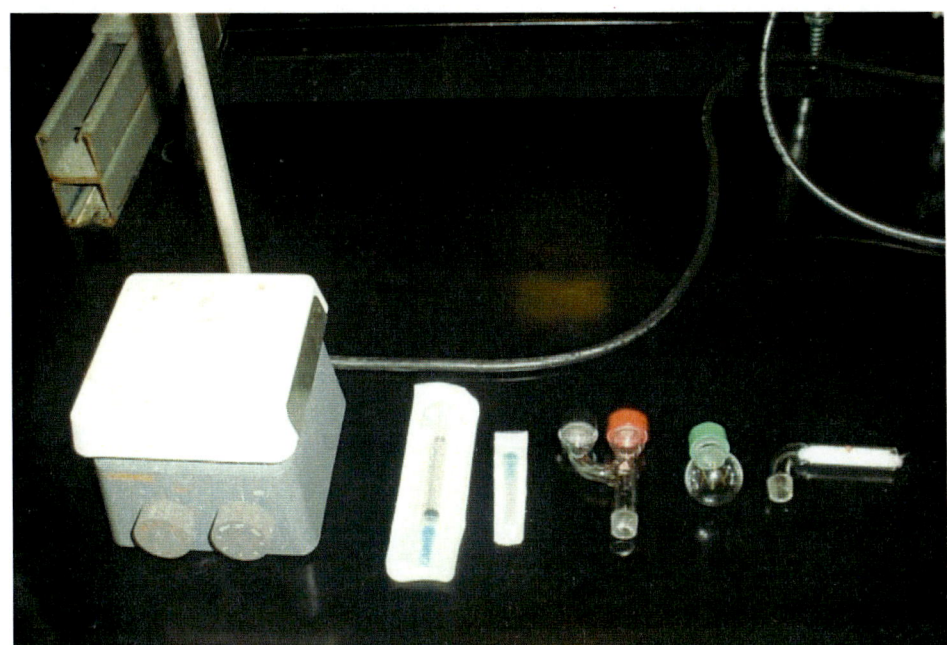

Materials Needed

Set-up

Benzocaine Experiment

Equipments Needed

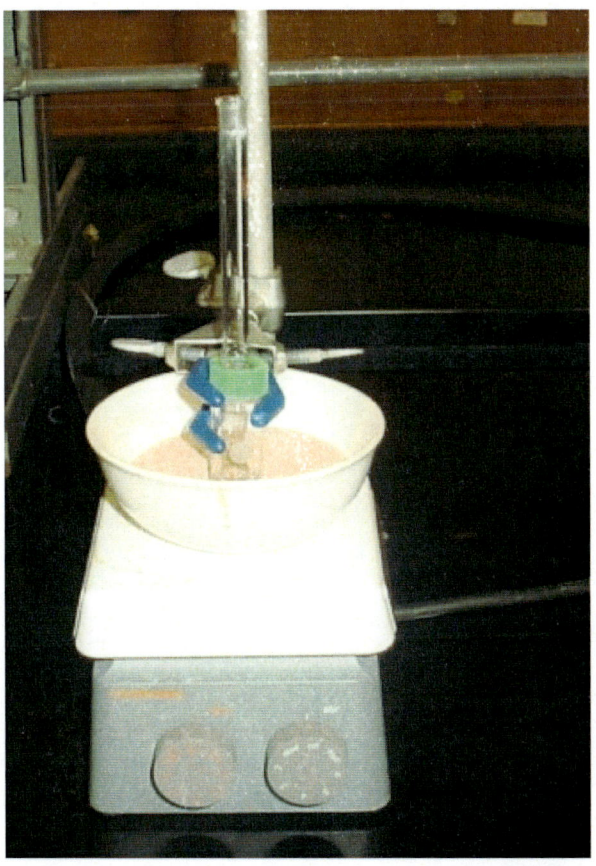

Set-up

CHEMISTRY CENTRAL LABORATORY
AT CAPITAL UNIVERSITY

COMMON ORGANIC SOLVENTS

Solvent	Boiling Point (⁰C)	Specific Gravity (g/ml)
Acetic Acid	118	1.05
Acetic Anhydride	140	1.08
Acetone	56	0.79
Benzene*	80	0.88
1-Butanol	118	0.81
Carbon Tetrachloride*	77	1.59
Chloroform*	61	1.48
Cyclohexane	81	0.78
ρ -Cymene	177	0.86
Diaxane*	101	1.03
Ethanol	78	0.80
Ether (Diethyl)	35	0.71
Ethyl Acetate	77	0.90
Hexane	69	0.66
Ligroin	60-90	0.68
Methanol	65	0.79
Methylene Chloride	40	1.32
Pentane	36	0.63
Petroleum Ether	30-60	0.63
1-Propanol	98	0.80
2-Propanol	82	0.79
Pyridine	115	0.98
Tetrahydrofuran	65	0.99
Toluene	111	0.87
m-Xylene	139	0.87

Solvents indicated in boldface are flammable.

*Suspect carcinogen.

Ref: VWR International, Chemicals for Every Application, poster (2000)

CONCENTRATED ACIDS AND BASES

Reagent	HCI	HNO$_3$	H$_2$SO$_4$	HCOOH	CH$_3$COOH	NH$_3$(NH$_4$OH)
Specific Gravity	1.18	1.41	1.84	1.20	1.06	0.90
% Acid or Base (by weight)	37.3	70.0	96.5	90.0	99.7	29.0
Molecular Weight	36.47	63.02	98.08	46.03	60.05	17.03
Molarity of Concentrated Acid or Base	12	16	18	23.4	17.5	15.3
Normality of Concentrated Acid, or Base	12	16	36	23.4	17.5	15.3
Volume of Concentrated Reagent Required to Prepare 1 litre of 1M Solution (ml)	83	64	56	42	58	65
Volume of Concentrated Reagent Required to Prepare 1 litre of 10% Solution (ml)*	227	101	56	93	95	384
Molarity of a 10% Solution	2.74	1.59	1.02	2.17	1.67	5.87

*Percent solutions by weight

Ref: VWR International, Chemicals for Every Application, poster (2000).

ORGANIC EQUIPMENT PICTURES

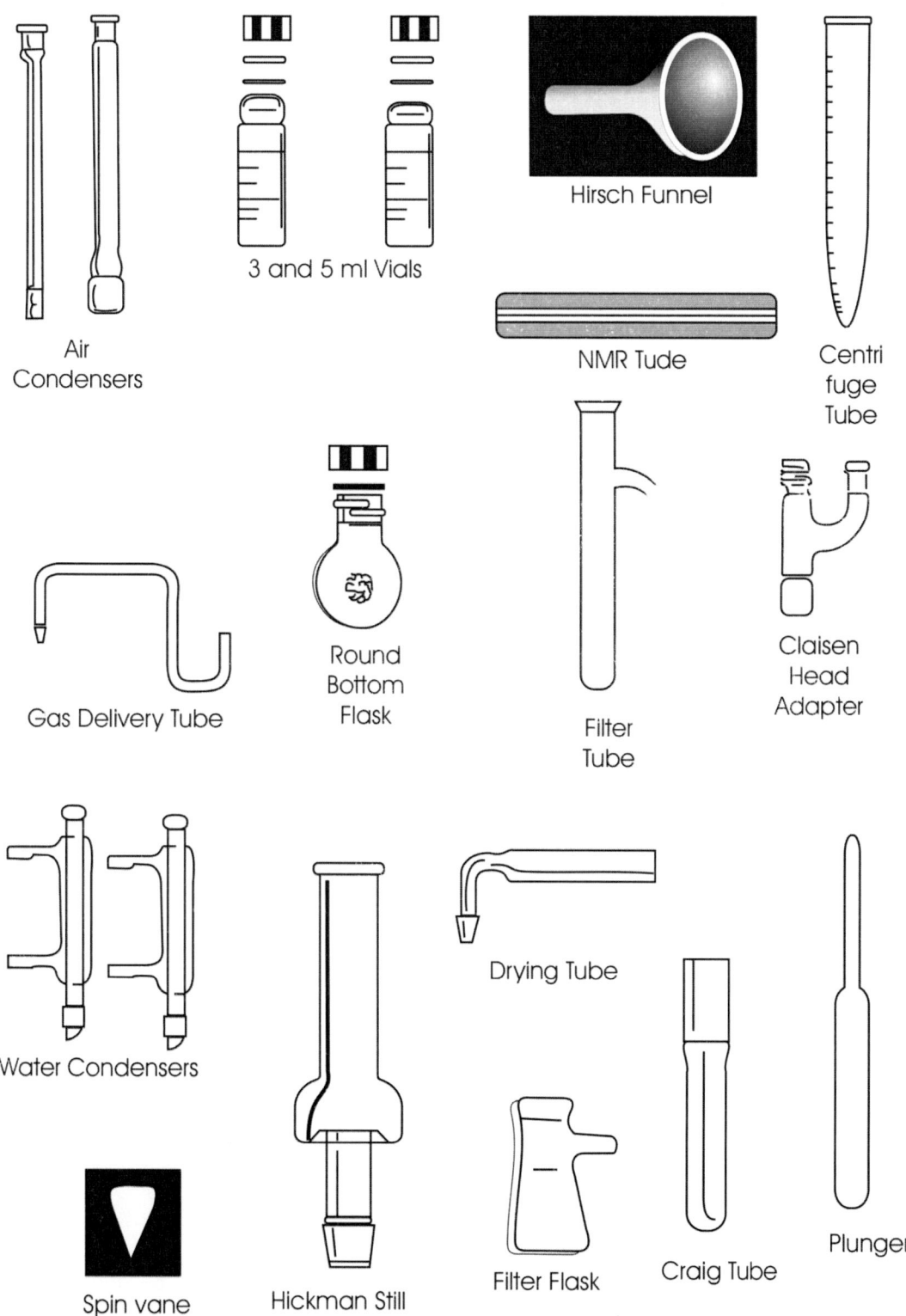

Air Condensers

3 and 5 ml Vials

Hirsch Funnel

NMR Tude

Centri fuge Tube

Gas Delivery Tube

Round Bottom Flask

Filter Tube

Claisen Head Adapter

Water Condensers

Spin vane

Hickman Still

Drying Tube

Filter Flask

Craig Tube

Plunger

Periodic Chart of the Elements

1 H 1.00794																	2 He 4.002602
3 Li 6.941	4 Be 9.012182											5 B 10.811	6 C 12.0107	7 N 14.00674	8 O 15.9994	9 F 18.9984032	10 Ne 20.1797
11 Na 22.989770	12 Mg 24.3050											13 Al 26.981538	14 Si 28.0855	15 P 30.973761	16 S 32.066	17 Cl 35.4527	18 Ar 39.948
19 K 39.0983	20 Ca 40.078	21 Sc 44.955910	22 Ti 47.867	23 V 50.9415	24 Cr 51.9961	25 Mn 54.938049	26 Fe 55.845	27 Co 58.933200	28 Ni 58.6934	29 Cu 63.546	30 Zn 65.39	31 Ga 69.723	32 Ge 72.61	33 As 74.92160	34 Se 78.96	35 Br 79.904	36 Kr 83.80
37 Rb 85.4678	38 Sr 87.62	39 Y 88.90585	40 Zr 91.224	41 Nb 92.90638	42 Mo 95.94	43 Tc (98)	44 Ru 101.07	45 Rh 102.90550	46 Pd 106.42	47 Ag 107.8682	48 Cd 112.411	49 In 114.818	50 Sn 118.710	51 Sb 121.760	52 Te 127.60	53 I 126.90447	54 Xe 131.29
55 Cs 132.90545	56 Ba 137.327	57 La 138.9055	72 Hf 178.49	73 Ta 180.9479	74 W 183.84	75 Re 186.207	76 Os 190.23	77 Ir 192.217	78 Pt 195.078	79 Au 196.96655	80 Hg 200.59	81 Tl 204.3833	82 Pb 207.2	83 Bi 208.98038	84 Po (209)	85 At (210)	86 Rn (222)
87 Fr (223)	88 Ra (226)	89 Ac (227)	104 Rf (261)	105 Db (262)	106 Sg (263)	107 Bh (262)	108 Hs (265)	109 Mt (266)	110 (269)	111 (272)	112 (277)		114 (289) (287)		116 (289)		118 (293)

58 Ce 140.116	59 Pr 140.90765	60 Nd 144.24	61 Pm (145)	62 Sm 150.36	63 Eu 151.964	64 Gd 157.25	65 Tb 158.92534	66 Dy 162.50	67 Ho 164.93032	68 Er 167.26	69 Tm 168.93421	70 Yb 173.04	71 Lu 174.967
90 Th 232.0381	91 Pa 231.03588	92 U 238.0289	93 Np (237)	94 Pu (244)	95 Am (243)	96 Cm (247)	97 Bk (247)	98 Cf (251)	99 Es (252)	100 Fm (257)	101 Md (258)	102 No (259)	103 Lr (262)

Useful Solvents

Useful Solvents

Solvent	M.M.	b.p.	dens.	f.p.	flam.	ε	P.I.	μ	n	S.W.	W.S.	Az.bp	%W	P.E.L.	Structure
acetic acid	60.05	118.0	1.049	16.7	x	6.15	6.2	1.70	1.3718	M	M	M	N	10	CH3COOH
acetone	58.08	56.3	0.790	-9.7	xx	20.7	5.1	2.69	1.3587	M	M	M	N	750	CH3COCH3
acetonitrile	41.05	81.6	0.782	-43.8	xx	37.5	5.8	3.44	1.3441	M	M	76.0	16.2	40	CH3CN
benzene	78.11	80.1	0.879	5.5	xxx	2.28	2.7	0	1.5011	0.18	0.06	69.3	8.8	1	C6H6
n-butanol	74.12	117.7	0.810	-88.6	x	15.8	3.9	1.75	1.3993	7.81	20.07	92.7	42.5	50	CH3(CH2)3OH
butanone	72.10	80.1	0.805	-86.7	xx	18.51	4.7	2.76	1.3788	24.0	10.0	73.4	11.3	200	CH3COCH2CH3
t-butyl methyl ether	88.15	55.2	0.711	-108.6	xxx					1.5	1.5		1.0		CH3OC(CH3)3
carbon tetrachloride	153.84	76.8	1.594	-23.0	0	2.24	1.6	0	1.4601	0.08	0.008	66.0	4.1	2	CCl4
chloroform	119.39	61.2	1.489	-63.6	0	4.81	4.1	1.15	1.4458	0.82	0.056	56.1	2.8	2	CHCl3
cyclohexane	84.16	80.7	0.779	6.5	xxx	2.02	0.2	0	1.4262	0.01	0.01	69.0	9.0	300	C6H12
diethyl ether	74.12	34.6	0.713	-117.1	xxx	4.34	2.8	1.15	1.3521	6.89	1.26	34.4	1.26	100	(CH3CH2)2O
dimethylformamide (DMF)	73.09	153.0	0.949	-60.4	x	36.7	6.4	3.86	1.4305	M	M			10	HCON(CH3)2
dimethylsulfoxide (DMSO)	78.13	189.0	1.100	18.5		46.7	7.2	3.9	1.4783	M	M			NL	CH3SOCH3
1,4-dioxane	88.10	101.3	1.034	11.8	x	2.25	4.8	0.45	1.4224	M	M	87.8	18.0	25	C4H8O2
ethanol	46.07	78.5	0.789	-114.1	xx	26.0	5.2	1.70	1.3610	M	M	78.2	4.0	1000	CH3CH2OH
ethyl acetate	88.10	77.1	0.901	-84.0	xx	6.02	4.4	1.88	1.3724	8.7	3.3	70.4	8.5	400	CH3(CH2)2OOCH3
n-hexane	86.17	68.7	0.659	-95.3	xxx	1.89	0.1	0.08	1.3749	0.001	0.01	61.6		50	CH3(CH2)4CH3
isopropanol	60.09	82.3	0.785	-88.0	x	19.92	3.9	1.66	1.3772	M	M	80.3	12.6	400	CH3CHOHCH3
methanol	32.04	64.7	0.791	-98.0	xx	32.7	5.1	1.70	1.3290	M	M	N	N	200	CH3OH
methylene chloride	84.94	39.8	1.326	-95.1	0	8.93	3.1	1.14	1.4241	1.6	0.24	38.1	1.5	50	CH2Cl2
petroleum ether		30-60	0.64-0.67	-40	xxx		0.1		1.36-1.38					NL	5-6 Carbon Hydrocarbon
tetrahydrofuran (THF)	72.10	66.0	0.888	-108.5	xx	7.6	4.0	1.75	1.4072	M	M			200	C4H8O
toluene	92.13	110.6	0.867	-95.0	xx	2.38	2.4	0.31	1.4969	0.053	0.033	84.1	13.5	50	PhCH3
water	18.02	100.0	0.998	0.0	0	80.1	10.2	1.87	1.3330					NL	H2O

M.M. molecular mass (g/mol), b.p. boiling point (C), dens. density (g/ml) at 20 C, f.p. freezing point (C), flam. flammability: x nonflammable, xx nonflammable but combustible, xxx very flammable, xxxx very highly flammable, ε dielectric constant at 20 C, P.I. polarity index of R. Snyder / Chromat. Sci. 16 223 (1978), μ dipole moment (D) at 25 C, n index of refraction at 20 C, S.W. solubility in water (%w/w) at 20 C, W.S. water solubility, solubility of water in solvent (%w/w) at 20 C, Az.bp azeotrope with water, boiling point (C), %W ... P.E.L. permissible exposure limit, the maximum concentration of airborne solvent (ppm) according to the American Conference of Governmental Industrial Hygienists (ACGIH), NL not listed.

Ref: VWR International, Chemicals for Every Application, poster (2000).

References

Sources of IR Correlation Chart, Correlation of 1H Chemical Shifts with Environment, Interpreted Spectra, and Problems.

1. Introduction to Spectroscopy, 3rd Edition, D. Pavia, G. Lampman, and G. Kriz, Harcourt College Publishers (2001).

2. Spectrometric Identification of Organic Compounds, 4th Edition, R.M. Silverstein, G.C. Bassler, and T.C. Morrill, Wiley & Sons, Inc. (1981).

3. Organic Structures from Spectra, S. Sternhell and J.R. Kalman, Wiley & Sons Inc. (1987)

4. I have no idea of where some of the spectra first originated.

Student Notes: